Returning Carbon to Nature: Coal, Carbon Capture, and Storage

Michael Stephenson

ELSEVIER

AMSTERDAM • BOSTON • HEIDELBERG • LONDON • NEW YORK • OXFORD
PARIS • SAN DIEGO • SAN FRANCISCO • SINGAPORE • SYDNEY • TOKYO

Elsevier
225 Wyman Street, Waltham, MA 02451, USA

First published 2013

Copyright © 2013 Elsevier Inc. All rights reserved

Notices
Knowledge and best practice in this field are constantly changing. As new research and experience broaden our understanding, changes in research methods, professional practices, or medical treatment may become necessary.

Practitioners and researchers must always rely on their own experience and knowledge in evaluating and using any information, methods, compounds, or experiments described herein. In using such information or methods they should be mindful of their own safety and the safety of others, including parties for whom they have a professional responsibility.

To the fullest extent of the law, neither the Publisher nor the authors, contributors, or editors, assume any liability for any injury and/or damage to persons or property as a matter of products liability, negligence or otherwise, or from any use or operation of any methods, products, instructions, or ideas contained in the material herein.

British Library Cataloguing-in-Publication Data
A catalogue record for this book is available from the British Library

Library of Congress Cataloging-in-Publication Data
A catalog record for this book is available from the Library of Congress

ISBN: 978-0-12-407671-6

For information on all Elsevier publications
visit our website at **store.elsevier.com**

This book has been manufactured using Print On Demand technology. Each copy is produced to order and is limited to black ink. The online version of this book will show color figures where appropriate.

Working together
to grow libraries in
developing countries

www.elsevier.com • www.bookaid.org

Printed and bound by CPI Group (UK) Ltd, Croydon, CR0 4YY

Transferred to digital print 2012

DEDICATION

To Jack and Fred

CONTENTS

From the window of the room where I write I can see a plume of condensing water that hangs over the south of Nottinghamshire day-in day-out. On some days it seems to ascend to a huge height and its upper surfaces are white and billow outwards like summer thunderclouds. Underneath is Ratcliffe-on-Soar power station.

Ratcliffe-on-Soar power station is one of the most efficient coal-fired power stations in Britain. It has four 500 MW turbines which produce enough electricity for 2 million houses burning coal that was formed in the Carboniferous period (approximately 300 million years ago). Some of it used to come from the Daw Mill coal mine about 40 miles away. The mine is situated in the small Warwickshire coalfield between the English towns of Birmingham, Nuneaton and Banbury, but the coalfield was once a swamp in an embayment in the ancient landmass called by geologists St. Georges Land.

As well as producing the electricity I need to power my house (and the laptop on which I'm writing), the power station also emits some $8-10$ million tonnes of CO_2 annually making it the eighteenth highest CO_2 emitting power station in Europe. For anybody with time on their hands they can work out that this is equivalent to 1000 tonnes CO_2 emitted per hour. For those of you with a liking for comparisons, Ratcliffe-on-Soar's production would take $6-9$ min to fill London's Albert Hall. Britain's private cars produce 91.0 million tonnes of CO_2 annually.

All the statistics about this power plant are spectacular: 48 million cubic metres of cooling water is taken from the nearby River Trent every year and of that some 11 million cubic metres of water are lost through the eight cooling towers. Its smoke stack is nearly 200 m tall, the tallest man-made structure in Nottinghamshire.

But I would not want to paint an ugly picture of the power station. The plant is compliant with the 'Large Combustion Plant Directive', an EU rule that aims to reduce acidification, ground-level ozone and particulates, by controlling the emissions of sulphur dioxide, oxides of nitrogen and dust from large combustion plants. To reduce emissions of sulphur the plant is fitted with flue gas desulphurization. This means that around 92% of the sulphur dioxide is safely removed before the flue gases are released into the environment. The plant also stands in grounds that are well tended and the station itself when you see it from the nearby M1 national motorway looks impressive rather than ugly. Sights like this are a necessary part of modern Britain, and any country where power is needed in very large quantities. Around the world there are 50,000 power stations like this (not all as big and not all burning coal); even in the United States alone there are 8000 power stations.

The central tension of modern energy and the main theme of this book is that we need energy, but we don't want the carbon dioxide emissions that come with it if we burn fossil fuels. This is because it's widely believed that rising CO_2 in the atmosphere is causing climate change.

But it's important to know that we've arrived at a partly coal-powered world through a long process that began in the industrial revolution — and before when coal formed for the first time in earth's history in large quantities in the Carboniferous period. The global circulation of carbon is known as the carbon cycle and this book will look at the carbon cycle in relation to coal formation, fossil fuel burning for electricity generation and climate change. The irony is that one of our best chances to keep emissions down is to bury the CO_2 back in the rocks where it came from, a technology that's called carbon capture and storage.

ACKNOWLEDGEMENTS

Thanks to Ruth O'Dell for very perceptive and thoughtful comments.

Thanks also to the BGS CCS Team without whose expertise this book would not have been possible.

Particular thanks to Andy Chadwick, Sam Holloway, Sarah Hannis, Michelle Bentham, John Williams, Ceri Vincent, Antony Benham, Andrew Bloodworth, Mike Ellis, Sev Kender, Chris Rochelle, Dave Jones, Derek Taylor, Gillian Dredge and Jonathan Pearce.

CHAPTER 1

Of Hockey Sticks and Coal

This chapter describes the rise of CO_2 in the atmosphere since the industrial revolution, and how we know that most of that rise is because of the burning of fossil fuels. In fact much of it is probably from coal. But why is human society so tied up with coal — why is the relationship so close? I'm old enough to remember as a boy sitting in front of a coal fire at home and thinking how much warmer it was than a wood fire. Even then it was obvious to me that burning coal makes a lot of heat. But it's not just that — coal is cheap and convenient. It was convenient at the beginning of the industrial revolution in what have now become rich countries, and it is still a convenient source of electricity in developing countries which will probably use a lot more of it in the future. Electricity powers development and drags people out of poverty and so coal is closely related with economic progress.

Keywords: *coal; climate change; industrial revolution; CO_2*

Most people have heard of the hockey stick graph in relation to global warming. The term was first used by the climatologist Jerry Mahlman who pointed out the shape of global temperature change through time was approximately flat up to about 1900 apart from the 'medieval warm period' and the 'little ice age' (the shaft of the stick), and then

Returning Carbon to Nature: Coal, Carbon Capture, and Storage. DOI: http://dx.doi.org/10.1016/B978-0-12-407671-6.00001-X

rapidly upturned into a 'blade' which got steeper with time. The hockey stick graph for CO_2 is rather more uniform in shape but broadly similar as might be expected since the chief control on temperature rise is believed to be the atmospheric content of CO_2. In this chapter I'll look at the shape of the CO_2 hockey stick graph and the part that coal had in creating it. I'll also look at the future for coal.

In June 2012 the concentration of CO_2 in the atmosphere reached 395 parts per million by volume (ppmv) or 0.000395%. It has been rising at about 2 ppmv every year since 2000. We can be absolutely sure of these amounts because since 1958 they have been measured directly from the atmosphere at Mauna Loa in Hawaii. Before that date we rely on more indirect measurements of bubbles of air trapped in cores of ice from Antarctica and Greenland. Antarctic cores show that atmospheric CO_2 levels were about 260–280 ppmv immediately before industrial emissions began in the mid-eighteenth century and did not vary much from this level during the preceding 10,000 years. Going back 400,000 years, which is possible by looking at progressively deeper and deeper ice, a pattern of CO_2 concentration is revealed which synchronizes with ice age cycles (Figure 1.1), but nowhere is there a sign of such a large leap of CO_2 concentration nor a rate of increase so extreme as in the last few decades.

To understand what part coal might have played in increasing CO_2 levels we need to know the fluxes of CO_2 in and out of the atmosphere, both natural and man-made. Volcanoes belch out CO_2 during eruptions

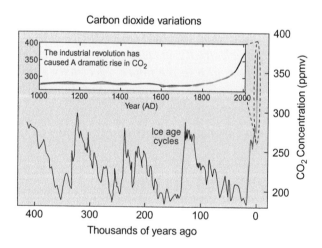

Figure 1.1 Carbon dioxide variation in the last 400,000 years. Inset shows the variation since 1000 AD. From Wikipedia.

but the natural burning of peat, wildfires, oxidation of organic matter in exposed rock and the respiration processes of living things, also all produce CO_2. Of course human beings breathe out CO_2. In fact we breathe out about 5% CO_2 or about 25 ml with each breath. Taking an average of 13,000 breaths a day an average person exhales about 1 kg of CO_2.

Man-made industrial sources of carbon dioxide include burning of fossil fuels for heating, power generation and transport, as well as processes such as cement and ammonia manufacture.

What about processes that take CO_2 out of the atmosphere? Plants convert carbon dioxide to carbohydrates during photosynthesis. They gain the energy needed for this reaction through the absorption of sunlight by pigments such as chlorophyll. The resulting gas, oxygen, is released into the atmosphere, and then used for respiration by animals and other plants, forming a cycle, which is part of the larger carbon cycle. Weathering of rocks containing silicate minerals also sucks CO_2 from the atmosphere converting it to insoluble and soluble carbonates.

The inputs and outputs of CO_2 to and from the atmosphere are nearly balanced. For example, the natural decay of organic material in forests and grasslands, and forest fires, releases about 439 gigatonnes (a gigatonne is a billion tonnes) of carbon dioxide every year, while new plant growth more than counteracts this effect, absorbing 450 gigatonnes per year. Although the carbon dioxide in the atmosphere of the ancient Earth shortly after its formation was produced by volcanic activity, modern volcanic activity releases only 130–230 megatonnes (million tonnes) of carbon dioxide each year, which is less than 1% of the amount released by human activities (at approximately 29 gigatonnes). In the pre-industrial era these fluxes were largely in balance. Now the balance is lost because of increased man-made CO_2 emissions: currently only about 57% of human-emitted CO_2 is removed by the biosphere and oceans, so man-made CO_2 is on the rise.

Direct evidence that this increase is from fossil fuel burning comes from the isotopic composition of the carbon in the CO_2 of the atmosphere through time (Figure 1.2). Carbon occurs naturally in the form of three isotopes ^{14}C, ^{13}C and ^{12}C, and the ratios of the different isotopes are quite significant. The graph shows the isotopic ratio of the isotopes ^{13}C : ^{12}C which is known as $\delta^{13}C$ and is expressed in parts per thousand using the symbol ‰ (the equivalent of a percentage but based on a thousand).

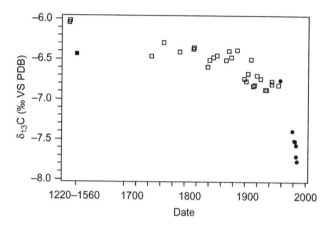

Figure 1.2 $\delta^{13}C$ records of the carbon in atmospheric CO_2 from Mauna Loa Observatory and ice core from South Pole ice. The trend is clearly decreasing. Circles from Mauna Loa Observatory and squares from South Pole ice. From Ghosh and Brand (2003).

The graph shows that in the last 200 years the $\delta^{13}C$ of CO_2 in the atmosphere has decreased by 1.5‰. Ice core data show the trend across a greater time period (from 1700 AD) with a particular increase in the steepness of the curve at about 1800, not long after the start of the industrial revolution.

So what's happened? It's partly a matter of dilution of atmospheric CO_2 with older carbon in CO_2 released from fossil fuels. The carbon in CO_2 that's released from burning coal, gas and oil has much lower $\delta^{13}C$ than atmospheric CO_2. It's old because it was incorporated into ancient plants millions of years ago and then formed into coal. For example the carbon in 300 million-year-old Carboniferous coal has a $\delta^{13}C$ of between -23 and -24‰ because this was the $\delta^{13}C$ in the plant tissues that formed the coal. This is much lower than the $\delta^{13}C$ of the CO_2 in the modern atmosphere (about -8‰), and so when the low $\delta^{13}C$ CO_2 of the burned Carboniferous coal enters the atmosphere, the $\delta^{13}C$ in the mix of modern atmospheric CO_2 decreases. The changing values of $\delta^{13}C$ through time as shown in the graph are also interesting because they in turn will be incorporated into modern plants as time goes by. In the next chapter I'll explain how this is useful in understanding the long-term carbon cycle.

More circumstantial evidence of the connection between fossil fuels and CO_2 in the atmosphere comes by simple comparison of historical data on fossil fuel burning and CO_2 concentration. This was shown by

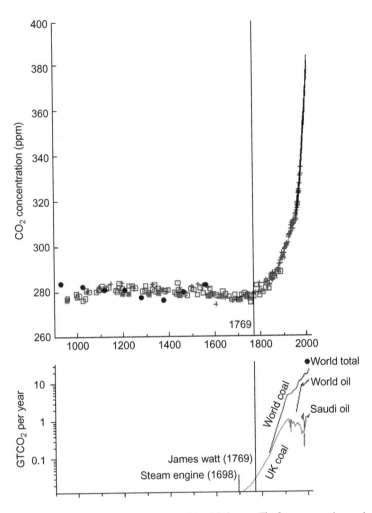

Figure 1.3 Atmospheric CO₂ concentration and historical fossil fuel usage. The first steam engine was invented in 1698, but James Watt patented his steam engine in 1769. From MacKay (2009).

David MacKay in his excellent book *Sustainable Energy without the Hot Air*. Mackay simply drew graphs of coal and oil burning on the same horizontal axis as for CO_2 concentration (Figure 1.3). In the vertical axis in his lower graph he used a rather unfamiliar unit of gigatonnes of CO_2 per year ($GtCO_2$ per year). Rather than the number of barrels or tonnes of oil and coal used, this is a figure that shows how much CO_2 was emitted from burning those barrels of oil or tonnes of coal.

The similarity of the curves is striking. The upper curve – the CO_2 hockey stick – has an exact counterpart in the lower curve. Also striking is the relationship with advances such as the development of the steam engine.

In this section we've seen that increasing CO_2 is a mainly a result of industrial activity. The smokestacks of power stations, cement factories and oil refineries are the smoking guns of the rapid CO_2 rise. In the next section I'll look at how human society became so attached to coal in the industrial revolution and explain why coal was such an integral part of it. I'll also show how the emerging economies of the eighteenth century were already reliant on coal.

COAL AND THE INDUSTRIAL REVOLUTION

Neither medieval coal miners nor later historians looking at the map of British coalfields operating in 1700 (Figure 1.4) would have been aware that their distribution is partly a consequence of the 300-million-old arrangement of swamps and subsiding land close to an emerging mountain range. This is something we'll examine later in this book. But the distribution of coal either at or near the surface in eighteenth century England affected the development of industry, and before that the development of trade in coal primarily for domestic heating.

Even in Roman Britain a small trade had developed along the North Sea coast supplying coal to London but by the middle of the sixteenth century supplies of wood for household heating were running out, particularly in the larger towns, and so the use of coal as a domestic fuel increased rapidly. By the beginning of the eighteenth century coal output in Britain was many times greater than that in the whole of the rest of the world.

Britain had a rising population – the population doubled between the early sixteenth century and the mid-seventeenth century – and many of these new people lived in towns far from the land that produced firewood. Firewood increased hugely in price with the result that coal very soon became the cheapest fuel, especially in towns along the east coast of England, including London, which could get coal by sea from the northeast. In the early seventeenth century much of Britain's dependence on coal was already established. By 1700, according to Hatcher (1993) fossil fuels had already 'eclipsed plant fuels as the leading provider of the nation's heat'.

In this early period water transport by canal or by sea was the only feasible method of moving large quantities of coal. All the large collieries were close to water, the biggest being in the northeast. By around 1700, 700,000 tonnes of coal then costing £125,000–£130,000 were shipped annually from the northeast.

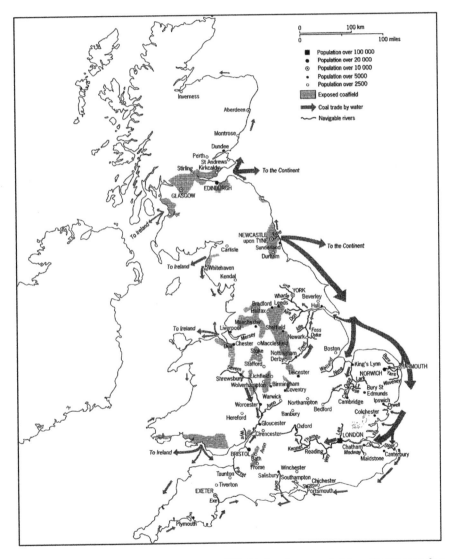

Figure 1.4 Coalfields and trade routes in Britain in 1700. Much coal was supplied down the east coast from Newcastle to the Midlands and London. From Hatcher (1993).

Things changed in the mid-eighteenth century. In 1769 James Watt patented his steam engine. This created a demand for coal apart from domestic heating, but also allowed mines to go deeper because the steam engines were connected to pumps that drained the deep mines below the water table. Before this, much less efficient chain pumps driven by water wheels drained the collieries. At deep levels *firedamp* (what geologists would now call coal bed methane) collected at great hazard to miners, and air circulation was originally achieved by lighting fires with obvious dangerous consequences. But steam engines allowed large fans to do the same work and so that ever more coal could be mined. Between 1769 and 1800, Britain's annual coal production doubled and its CO_2 emissions from coal burning were already close to 100 million tonnes of CO_2 per year.

Coal use on an industrial scale spread rapidly. In Germany mining on a large scale began in the 1750s in the Ruhr and Wurm valleys; similar iron and steel development occurred in Wallonia in Belgium. Further east in Poland, Silesia developed coal mining at a rapid pace, so that Poland has become a major world coal producer today, and one of the western nations most dependent on coal, particularly for electricity generation. In the northeast of the United States coal had become the main domestic fuel of the cities by 1850.

Coal caused problems long before anyone was concerned about CO_2 emissions. The Great Smog of London between December 1952 and March 1953 was caused by a combination of factors including a cold snap, fog and an increase in diesel powered buses on London's roads. The cold fog of the London winter trapped gases and particulates from the thousands of domestic coal fires and coal-fired factories so that they concentrated at ground level. The smoke got indoors and into public places causing great disruption. Apparently Londoners took the condition of the air stoically but it was soon found that the fog had killed 4000 people – most of whom were very young or elderly or had respiratory problems. Another 8000 died in the weeks and months that followed. The Clean Air Act of 1956 introduced rules to reduce air pollution, for example 'smoke control areas' in some towns and cities where only smokeless fuels could be burnt. It encouraged householders to use smokeless coal, electricity, and gas for heat and therefore reduced the amount of smoke pollution and sulphur dioxide from household fires. It also encouraged power stations to be sited away from cities, and for the height of some chimneys to be increased. The effects of the Act were almost immediate with levels of SO_2 and smoke falling throughout the 1960s and 1970s. The Great Smog helped to develop the modern environmental movement and shows how coordinated and rapid action was able to confront the problem. But the battle for better air quality continues in many fast growing cities. Before the Summer 2008 Olympic Games began China started its campaign of 'Defending the Blue Sky' to improve the poor quality of Beijing's air. This involved exiling more than 1000 heavy industrial and power-generation plants outside the city and the introduction of natural gas for some domestic and industrial use, as well as cleaner coal with reduced sulphur – with almost immediate effect on Beijing's atmosphere.

The problem of acid rain was noticed much later and is caused by sulphur in coal being oxidized to sulphur dioxide during burning. The SO_2 and SO_3 formed dissolve in rain water to make it acidic. Many power plants are fitted with flue gas desulphurization (FGD) which strips out the SO_2 from exhaust (flue) gas. But where there is no FGD, tall smokestacks designed to reduce local pollution simply spread acid rain further by releasing gases into regional atmospheric circulation. Rain with low pH may fall a long way downwind of the power station, and high rainfall mountainous areas suffer particularly.

The massive increase in CO_2 is just the latest pollution scare to come from burning coal. It might be the most subtle in the sense that we can't see CO_2 and it doesn't have an immediate health effect, but its result – irreversible global climate change – is perhaps the most serious.

WHERE IS THE COAL?

Most of the coal that powered the industrial revolution was formed in a period of intense coal sedimentation in the late part of the Carboniferous period. In Britain the Carboniferous-aged rocks that contain large amounts of coal are called the Coal Measures, and they date from about 320 to 300 million years ago. It's no accident that coal is similar across northeastern United States, Britain, Netherlands, Belgium, Germany and Poland because it was in these areas (which were joined together in the Carboniferous) that large tracts of subsiding land connected with mountain building were created; and newly evolved land plants colonized these low lying areas. These so-called *coal swamps* were continuous across what is now the eastern United States into eastern Europe. Their footprint is still visible – even though it's slightly modified by later erosion and mountain building we can still see the ancient geography of the Carboniferous through the filter of time. The map (Figure 1.5) also shows these early exploited coalfields and their influence on wealth – even now early in the twenty-first century, much of the world's wealth sits on the surface above Carboniferous-aged coalfields.

The later development of coal-fuelled industry in the developing world and in Australia used coal of a different age, most of it being younger, having been formed in forests rather different in character in the Permian period. The coal was formed by a probable deciduous seed plant rather than an evergreen clubmoss plant (as in the case of

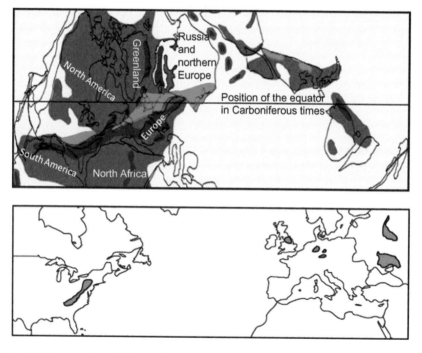

Figure 1.5 Coal as it was formed in Carboniferous swamps and as it appears today. Top – the arrangement of the continents during the period of the maximum Carboniferous coal formation between 320 and 300 million years ago, from Cleal and Thomas (2005). The land areas are shown in brown with the positions of modern continents marked. The coal was formed in almost continuous swathes of subsiding land populated by dense swamp forests (green). Bottom – the main coalfields of the eastern United States and Europe with modern continental positions restored. These were the first coalfields to be exploited in the industrial revolution and are the remnants of forested subsiding land after mountain building and erosion.

the Carboniferous). Reflecting these differences Permian Indian, South African and Australian coal occurs in layers (seams) of great thickness (30 m or more – much thicker than European Carboniferous coal seams) that extend over vast areas.

From a British perspective, the age of coal would seem to have passed, and British readers of newspapers would be forgiven for thinking that 'King Coal' is dead. Of course this isn't true globally; more coal is being used every year in power stations and steel making. A lot of coal is now being turned directly into liquid fuels close to mines (the so-called coal-to-liquid or CTL). Many countries with very remote coal resources, for example China's coal resources in inner Mongolia, regard CTL as a way of realizing the value of coal that would be too difficult and expensive to export and can be turned into something that you can put in a truck or a car.

Coal usage is continuing to rise and today's world uses nearly 5000 million tonnes per year (IEA, 2011). In 1980 the figure was 2500 million tonnes. Coal accounted for nearly half of the increase in global energy use over the past decade, driven mainly by surging demand in China's power stations and factories. Coal continues to be the second largest primary fuel globally and the backbone of world electricity generation. It is also the most abundant fossil fuel globally with reserves totalling 1 trillion tonnes or some 150 years worth of current production. But this isn't the only reason why coal is likely to be on the scene for many years into the future.

USE OF COAL IN THE FUTURE

Forecasting the demand and supply of coal and therefore its price on world markets is a difficult job. Despite its reputation for being dirty and dangerous to mine, coal still has fascination for many governments and industries. Coal is calorific and so packs an energy punch. As seen in the last section it is very abundant, and if you don't have transport it far, it's cheap.

The International Energy Agency (IEA) was established by the Organisation for Economic Co-operation and Development (OECD) in 1974 after the 1973 oil crisis. The IEA publishes a World Energy Outlook annually and the 2011 edition contains a whole section on coal, so important does it see this part of the energy market. In past years the World

Energy Outlook has predicted the future by means of 'scenarios' which are a bit like the alternative futures of science fiction stories. The scenarios contain predictions of energy infrastructure investment, energy demand and supply. The 2011 World Energy Outlook presents three scenarios, based primarily on the need to reduce CO_2 emissions and on common assumptions of economic conditions and population growth. They differ mainly in how various government policies might be applied. The scenarios factor in the concept of 'lock in' – the natural inertia that large energy installations like power stations have. Coal power stations typically have lives of 40 or 50 years, and once they are built it's difficult and expensive to change them.

The three IEA scenarios are the 450 Scenario, the Current Policies Scenario and the New Policies Scenario. In terms of limiting CO_2 emissions, the 450 Scenario is the most stringent in that its projections are constrained by the need to limit concentration of greenhouse gases in the atmosphere to around 450 parts per million. This concentration of CO_2 is widely believed to be required to limit global increase in temperature to $2°C$. This is the 'wish list' scenario that perhaps might never be achievable.

On the other side of the coin is the Current Policies Scenario which looks at present policies in energy assuming that they will not change and calculates the future from this standpoint regardless of the climate consequences. It's similar to the *business as usual* projections in climate change that often predict ruinous change to the environment within the next century.

The third is the New Policies Scenario which takes account of broad policy commitments and plans that have been announced by countries and their governments, including national pledges to reduce greenhouse gas emissions and plans to phase out fossil energy subsidies, even if the measures to implement these commitments have yet to be identified or announced. This might be regarded as the most realistic of the scenarios if you are an optimist!

But what do these scenarios predict for coal usage? Taking the text directly from the IEA (2011) World Energy Outlook: 'In the New Policies Scenario, global coal use rises through the early 2020s and then remains broadly flat, above 5 850 million tonnes of coal equivalent (Mtce), through to 2035 – one-quarter higher than in 2009. Coal remains

the second-largest primary fuel and the backbone of electricity generation. In the Current Policies Scenario, demand carries on rising after
2020, increasing overall by nearly two-thirds to 2035. But in the 450
Scenario, coal demand peaks before 2020 and then falls heavily, declining
one-third between 2009 and 2035'.

The 450 Scenario illustrates the polluting power of coal because its
carbon-constrained core needs to limit coal above most other fuels.
But perhaps the New Policies Scenario most clearly shows the way
ahead. It shows rising coal use into the 2020s and then indicates a pla-
teau after that. It also remains the main way that world will generate
electricity. So coal is here for a while longer.

The most sensible way of proceeding is to work towards the 450
Scenario but the New Policies Scenario seems more likely. But why is it
more likely? Why is the world hooked on coal when it knows that it's
bad for its health? In the next section I'll consider three countries –
India, China and South Africa – where coal has a clear role in the
future and where coal has a special significance beyond mere electricity.
Later in this book I'll return to these countries to look at how they
might be able to mitigate the climate changing potential of burning coal
by carrying out carbon capture and storage.

Coal in India
In India coal is equated – through rural electrification – to poverty
alleviation and health. India has a third of the world's poor. In 2011,
33% of Indians fell below the international poverty line of US$ 1.25
per day, while 69% of Indians live on less than US$ 2 per day.
Amongst these poor Indians, who live mostly in rural areas, domestic
use of electricity is rare. India alone accounts for more than 35% of
the world's population without electricity access. Most domestic energy
in India comes from biomass (firewood, crop residue and dung), and
India consumes 200 Mtoe (million tonnes of oil equivalent) of biomass
each year (Figure 1.6). One hundred million Indian households still
use firewood to cook food, mainly in rural areas.

All this cooking with firewood takes its toll on the health of Indians
with an estimated 50,000 deaths per year (household fires, accidents and
ill health). Worldwide, exposure to smoke emissions from the household
use of solid fuels is estimated to result in 1.6 million deaths annually

(World Bank 2003). Illnesses are also caused by lighting using kerosene lamps — which is the most common type of lighting in rural India (Figure 1.6). So the aim of India's rural electrification programme is clear: electricity is much healthier in households. But another aim of the programme is to improve agricultural production (through for example better irrigation pumps) and to develop business and trade in agriculture. Partly due to successful electrification, India adds 40 million people to its middle class every year — and also through irrigation has made the country self-sufficient in food grain production.

Where will India's new electricity come from? Most will come from coal. At present coal provides about 70% of India's electricity. But a record 20 GW was added in the year 2011–2012 bringing the electricity generation capacity of India in 2012 up to 200 GW. Renewable electricity provides over 12% of this capacity, and future rural electrification in remote areas is most likely to be off-grid solar lighting systems, irrigation pumps, biogas plants, solar cookers, biomass gasifiers and improved cooking stoves. But much rural electrification will be on the grid, and most of this will be provided by coal. The share of coal-based generation for India's future is planned to be around 80%.

A taste of this future is offered by the so-called Ultra Mega Power projects (UMPP). As you might imagine these are a series of 16 very large coal power plants of 4 GW or bigger — so twice the size of Ratcliffe-on-Soar power station. So far two of these massive plants are under construction in Madhya Pradesh and Gujarat, and permission has been granted for another two. Because of their huge size, the cost of electricity should be lower due to the economies of scale.

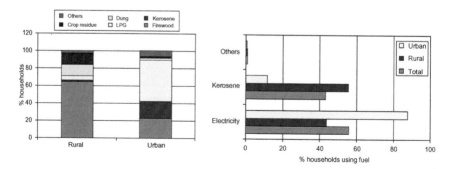

Figure 1.6 Domestic use of fuel in India. Left — urban—rural differences in energy for cooking in India. Right — distribution of Indian households by source of lighting. From Bhattacharyya (2006).

India has very large coal reserves but despite the development of UMPPs, it sometimes has electricity-supply shortfalls of up to 15% during peak demand hours and has this year (2012) suffered catastrophic collapse of electricity supply over periods of several days. Problems in the speed of deployment of UMPPs and in coal production have partly been to blame and have increased the dependence of Indian generating companies on imported coal. Nevertheless the Indian government is committed to coal and plans to improve mining technology and manpower to achieve its goal of rural electrification.

How does India's coal use change according to the IEA's scenarios that were described earlier? In the New Policies Scenario, India will become the world's second largest consumer of coal (after China) by around 2025, with demand almost doubling to 880 million tonnes per year by 2035 (Figure 1.7).

Although today, 300 million Indians lack domestic electricity, by 2030 this will have reduced by half. Also according to the New Policies Scenario, the demand for coal in industrial use will continue to rise averaging 4% per year growth up to 2035, mainly through the increased manufacture of crude steel. India is also a large cement manufacturer, and much industrial coal is used in the process. There is little doubt that this will grow too as more economic growth further spurs building.

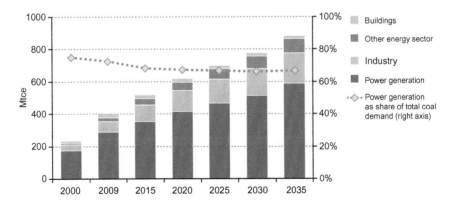

Figure 1.7 Coal demand in India by sector in the New Policies Scenario. Electricity will be the major user as far ahead as 2035. World Energy Outlook 2011 © OECD/IEA 2011, fig. 10.23, p. 388.

Coal in China

China is by far the largest coal consumer in the world, accounting for almost half of global coal use in 2010. Between 2005 and 2010 Chinese coal consumption increased by 50% spurred on by demand from electricity generators and industry. In the IEA New Policies Scenario, China's coal demand will increase by 30% to over 2850 million tonnes per year by 2020 and stabilize above 2800 million tonnes until 2035 (Figure 1.8). Coal will continue to provide more than half of China's electricity until 2035, even though China's 12th Five-Year Plan (2011–2015) sets targets for reducing energy use and CO_2 emissions, aiming to increase non-fossil fuel energy from 8.3% in 2010 to 11.4% in 2015.

China is likely now the world's largest electricity generator (having overtaken the United States this year, 2012), with about 1000 GW. The demand for electricity in China is not so closely related to rural electrification although the country has one of the largest such programmes in the world and foresees full rural electrification by 2015.

Most of the demand will continue to come from industry and manufacturing. There is some uncertainty over the future role of gas in electricity generation which may not have been sufficiently factored into the IEA's New Policies Scenario, and that is the growth of shale exploration gas in China. This began substantially in 2010 and which has not yet lead to significant production. If a 'dash for gas' happens in China as it did in the United States between 2005 and 2010 then

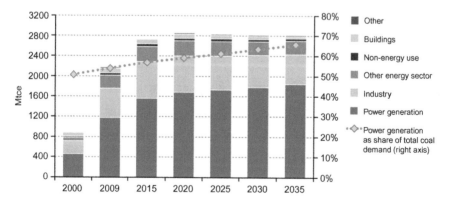

Figure 1.8 Coal demand in China by sector in the New Policies Scenario. Like India, China will use most of its coal to generate electricity. World Energy Outlook 2011 © OECD/IEA 2011, fig. 10.18, p. 382.

electricity generation might switch substantially to gas power (Figure 1.9). Other uncertainty hangs over the development of CTLs in China. As I mentioned earlier, China has a lot of coal in remote areas, for example in Inner Mongolia. Here Shenhua, one of China's giant coal companies, has opened a plant that generates liquid fuels from coal. Incidentally this is the home of one of the few active pilot carbon storage projects in China where a portion of the CO_2 from the CTLs plant is directly injected underground. The IEA estimates that Chinese CTLs will increase 10-fold by 2035, though admits that its feasibility depends on a high oil price.

Coal in South Africa

South Africa falls between India and China in that it has a large number of rural people without access to electricity (roughly 60% of South African households), but also a strong demand for electricity particularly for the mining industry. South Africa's coal reserves are large – 28 billion tonnes – which would allow 100 more years of mining at current rates. South Africa exports coal mainly through the port at Richard's Bay, but constraints on rail transport have reduced its capacity to bring coal from the interior and so export has become less important to the country. In fact coal supply to indigenous power stations has sometimes been inadequate to satisfy electricity demand in recent years with the result that the country has suffered power cuts in recent years.

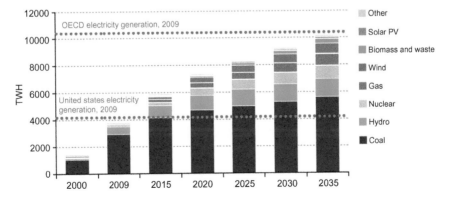

Figure 1.9 Electricity generation in China by type in the New Policies Scenario. World Energy Outlook 2011 © OECD/IEA 2011, fig. 10.21, p. 385.

According to the IEA New Policies Scenario, South African coal production will be driven mainly by domestic demand for coal for power supply. At present more than 90% of electricity is generated by coal in South Africa and this will remain the case well into the next decades. Coal production is predicted to rise to a peak of around 230 Mtce around 2020 and then fall to 210 Mtce by 2035 (Figure 1.10).

Two key points differentiate South Africa from India and China. One is the metals mining industry which is a big user of electricity particularly in the processing of ore. This industry employs almost half a million South African workers and generates almost a fifth of South Africa's GDP. In the last few years some mining operations have not been able to get enough electricity and this has had a considerable impact on investor confidence. Unusually South Africa also uses a lot of domestic coal in CTLs. Sasol the biggest CTL company in the world produces 36% of liquid fuels consumed in South Africa, mainly from coal.

Solving the Problem of Coal Use

The preceding sections show the enthusiasm with which some countries will use coal in the future. This is because the health of their people, control of poverty and economic development are all closely related to coal through electricity. Although I've chosen India, China and South Africa to illustrate this point, it's true of many developing nations, particularly those which are fortunate in having a lot of coal. Countries that have coal are likely to use it.

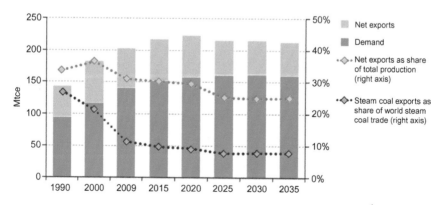

Figure 1.10 Forecast coal production, exports and demand in South Africa according to the IEA New Policies Scenario. World Energy Outlook 2011 © OECD/IEA 2011, fig. 11.25, p. 441.

All of this looks rather bleak for climate change, if coal powered electricity generation is unabated by for example CCS. The goal of enormous increases in coal powered electricity will mean more CO_2 emissions. But looked at another way — elevating more and more people into the middle classes — also has its problems because middle class people use much more energy.

Earlier I mentioned that India adds 40 million people to its middle class every year. What effect does this have? According to Grunewald (2012) the carbon footprint of high-income Indian households (1.2 tonnes of CO_2 per year per household) is about six times as high as the carbon footprint of poor households. Under ambitious poverty reduction targets in India, the annual growth rate of CO_2 emissions increases from 4.8% to 5.9%.

And so poverty alleviation implies carbon emission increases. No one would seriously suggest that poverty alleviation should stop. Can we allow poverty alleviation to continue in the developing world if the developed world really works hard at reducing its emissions? Is there enough headroom?

Let's look at the situation across the world. The most developed (the richest) countries emit the most CO_2 per person for obvious reasons: Europe's per-person greenhouse gas emissions are twice the world's average while North America's are four times the world's average. China — although it recently surpassed the United States as the world's largest CO_2 emitter — has per-person emissions less than the global average. India's emissions are less than half the world average.

Historical cumulative emissions are also much larger in the richer countries (it's one reason why they're rich!). Predictably, the 'old coal' nations of the industrial revolution, the United States, Britain and Germany are amongst the countries that top the historical cumulative emissions chart (Figure 1.11). Expressed as an average emission rate over the period 1880–2004, the United States for example had a rate of almost 10 tonnes of CO_2 per person per year.

Even if we look at a forecast of 'cumulative emissions' by 2035 the responsibility for emissions will still sit mainly with the historically rich nations (Figure 1.12).

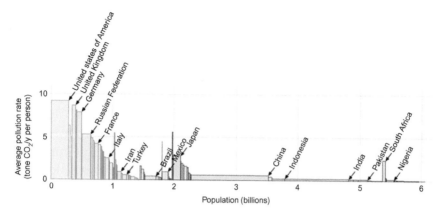

Figure 1.11 Historical emissions of nations between 1880 and 2004 (average per person per year). The width of the rectangles representing countries shows the sizes of their populations. From MacKay (2009).

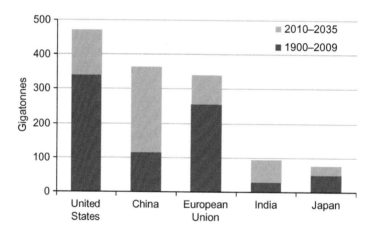

Figure 1.12 Cumulative and forecast CO$_2$ emissions. In China by 2035, cumulative CO$_2$ emissions between 2010 and 2035 will be larger than the emissions between 1900 and 2009. Presentation given by Dr. Fatih Birol © OECD/IEA 2011.

Depending on which scenario of emissions and climate change you use, a headroom or safe emissions 'space' can be calculated which shows the amount of freedom a country might have to continue to emit CO$_2$ within a framework of overall emissions reduction. This is the slack in the system. The space could be allocated among nations, industries or consumers. Perhaps industrialized economies should take the lead, given they have both the greatest resources available and the most historical responsibility. So they would have to decarbonize their energy systems early and allow late emitters the right to use up the safe 'emissions space'. Of course this relies on having very accurate figures that can be agreed upon internationally. I'll examine this in more detail later in this book when I consider the practical difficulties of 'accounting for carbon'.

This chapter has shown that countries set on a high coal use trajectory are unlikely to change their plans. A CO_2 emissions strategy for these countries therefore relies on an abatement method that is consistent with long-term coal use or coal use as a bridge to renewables. Such an abatement method could be carbon capture and storage.

BIBLIOGRAPHY

Anderson, K., Bows, A., 2011. Beyond 'dangerous' climate change: emission scenarios for a new world. Phil. Trans. R. Soc. A 369, 20–44.

Bhattacharyya, S.C., 2006. Energy access problem of the poor in India: is rural electrification a remedy? Energy Policy 34, 3387–3397.

Cleal, C.J., Thomas, B.A., 2005. Palaeozoic tropical rainforests and their effect on global climates: is the past the key to the present? Geobiology 3, 13–31.

Ghosh, P., Brand, W.A., 2003. Stable isotope ratio mass spectrometry in global climate change research. Int. J. Mass Spectrom. 228, 1–33.

Grunewald, N. 2012. The carbon footprint of Indian households. Paper Prepared for the 32nd General Conference of the International Association for Research in Income and Wealth Boston, USA, August 5–11, 2012.

Hatcher, J., 1993. The History of the British Coal Industry. Clarendon Press, Oxford.

IEA, 2011. World Energy Outlook, Paris.

MacKay, D., 2009. Sustainable Energy Without the Hot Air. UIT.

Metz, B., Davidson, O., de Coninck, H., Loos, M., Meyer, L. (Eds.), 2005. Carbon Dioxide Capture and Storage IPCC. Cambridge University Press, UK, pp 431.

Murthy, N.S., Panda, M., Parikh, J., 1997. Economic growth, energy demand and carbon dioxide emissions in India: 1990–2020. Environ. Dev. Econ. 2, 173–193.

Rural Electrification (India) Corporation 43rd Annual Report, 2012.

CHAPTER 2

The Negative Greenhouse

In this chapter I'll look at the formation of coal in the strange swamp forests of the Carboniferous Period 300 million years ago and follow the chain of unlikely coincidence that allowed this hard black concentrated carbon to form. The environment of the swamp forests was rather different from now with insects with metre wingspans, trees that grew 5 m in a year that photosynthesized through their bark and an oxygen rich atmosphere that made wildfires very common. Forests were a new 'invention' and their ability to capture carbon through photosynthesis was responsible for the formation of coal. But this changed the atmosphere sucking CO_2 out and storing it in coal — and probably changed the climate too.

Keywords: *fossils; coal; peat; climate change; carbon cycle; CO_2*

My local power station, Ratcliffe-on-Soar, delivers baseload electricity day-in day-out to the British electricity National Grid. To do this it has a close relationship with coal mines in the area, particularly the Daw Mill mine in Warwickshire. The mine was, until recently, one of the most productive in Britain and the last surviving mine in the Warwickshire Coalfield that once had 20 operating collieries.

The coalfield is shaped like an elongated triangle around 30 km wide at its widest and 50 km long and contains sediments of Westphalian age

Returning Carbon to Nature: Coal, Carbon Capture, and Storage. DOI: http://dx.doi.org/10.1016/B978-0-12-407671-6.00002-1

(the Westphalian is a subdivision of the Carboniferous Period). To the west, the coal seams disappear against a large fault and to the east the seams get thinner and less economic. The most important seam, and the one mined at Daw Mill is the 'Warwickshire Thick Coal', which reaches thicknesses of up to 8.5 m and is high in quality with little sulphur and few rock impurities (known as ash), though sometimes the seam contains layers of less organic-rich shale or sandstone which makes the coal less pure. The coal contains plant fossils — recognizable leaves, but also a myriad of fossil spores — that indicate that the coal was made in a forest of trees rather unlike those of modern forests. The forest itself grew in an ancient embayment of about 100 km² in a long-disappeared area of upland known as the Wales-Brabant Massif that stretched across what is now Wales, eastern England and into Belgium (Figure 2.1). To the north and south of this upland area were wide low-land swamps colonized by fast growing trees related to our modern club mosses, but which grew to heights of 30 m or more.

But Britain's share of the coal swamps of the Carboniferous is quite small. They stretched across northeastern America, northern Europe, eastern Europe, and into Russia and Kazakhstan. At their height they covered more than 2 million square kilometres (the area of modern day Argentina) and generated more of the world's hard black coal than at any other time, changing the modern world forever. What's not known so widely is that they also had an enormous effect on the world's climate 300 million years ago and may have helped to initiate one of the major glacial periods of geological history. These swamps were responsible for Britain's Carboniferous coal deposits but their existence is due to a set of rather extraordinary circumstances of plant evolution and mountain building. If things had been slightly different, Britain might never have inherited its valuable coal.

THE PLANTS

The Carboniferous lowland forest was dominated by five major plant groups, lycopsids (club mosses), sphenopsids (horsetails), filicopsids (ferns), pteridosperms (seed ferns — a slightly more evolved form of a fern) and cordaites (rather like a conifer). The first three reproduced with simple spores, while the last two used seeds. All five of these types produced large trees but also smaller bushes and shrubs (Figure 2.2). These are the plants that formed Carboniferous coal.

But what is interesting here is that swamp forests and dryland forests were a new feature of the land at this remote time in Earth's past. Although plants were well established on some parts of the land, large areas were probably still un-colonized by plants. To understand why ancient forests like these were so different to modern forests we have to understand a little of how land plants evolved.

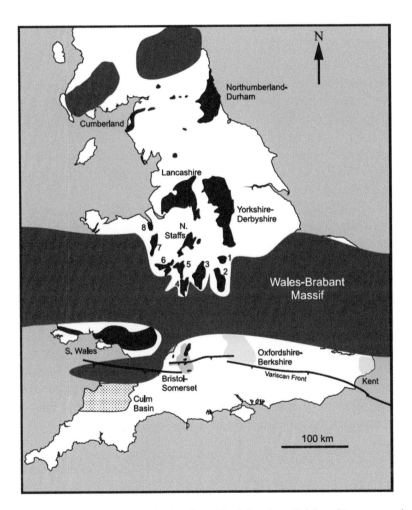

Figure 2.1 The main coalfields in England and Wales in Westphalian times which formed in swamps north and south of the Wales-Brabant upland area. Exposed coalfields shown in black, concealed coalfields in grey, areas of non-deposition in brown; areas of marine deposition in fine stipple. For coalfields immediately north of the Wales-Brabant Massif: 1, Leicestershire − S. Derbysire; 2, Warwickshire; 3, S. Staffordshire; 4, Wyre Forest; 5, Coalbrookdale; 6, Shrewsbury; 7, Denbigh; 8, Flint. From Cleal et al. (2010).

Figure 2.2 Carboniferous peat-forming plants: L, lycopsids (club mosses); S, sphenopsids (horsetails); F, filicopsids (ferns); P, pteridosperms (seed ferns – a slightly more evolved form of a fern) and C, cordaites (a type of very early coniferophyte). From Pfefferkorn et al. (2008).

Fossil evidence for the earliest land plants is limited to spores from moss or hornwort-like plants 475 million years old, 175 million years before the coal swamp forests. At this time there was no sign of the mysterious plants that produced the spores, and there was no sign of them until 50 million years later! The plants, when they eventually do appear as fossils, show a simple vascular system in that they had tissues for conducting water, minerals and photosynthetic products through the plant. This also meant that they had a certain rigidity (from vascular pressure) and so could reach up for sunlight, though not very far. The earliest known of these plants *Cooksonia* (mostly from the northern hemisphere) and *Baragwanathia* (from Australia) were only very small, millimetres high at the most. By 400 million years ago primitive plants had created the first recognizable soils that harboured mites and scorpions. Strangely these plants didn't have leaves; small leafless shrubs filled the landscape until the tree-like fern *Archaeopteris* appeared with the first leaves. By the start of the Carboniferous period 360 million years ago where all the action of this chapter takes place, plants and trees with leaves were common and the first seed-forming plants had appeared. For our story, the next most important event was the appearance of lignin.

Lignin (from the Latin word *lignum* meaning wood) is one of the most abundant organic polymers on Earth, exceeded only by cellulose. At the beginning of the Carboniferous it appears that a number of plant types were able to make lignin. Its main function was to strengthen wood (or xylem cells) in the stems of trees so that they could grow taller – at least taller than their competitors. Vascular pressure is not sufficient to allow trees to grow very tall. This 'invention' not only made trees taller and therefore more efficient photosynthesizers but also made them more *difficult to digest.*

Let me explain this. The lignin in trees — being a large complex molecule — is much more difficult to degrade than other plant tissues and is physically stronger too. When Carboniferous trees and plants with lignin died, their lignin tissues persisted longer than other tissues, remaining solid and un-rotted even in swamp waters. It may be that early primitive Carboniferous microbial scavengers, fungi and bacteria weren't able to break the lignin down either. A lot of carbon in trees is in lignin, and so if the trees died and were preserved in peat quickly, most of the carbon was buried. A large coal-forming Westphalian tree might have contained 3200 kg of carbon, mostly in its lignin, and if this was buried then most of the carbon was too. If we think of this happening over a vast area of 2 million square kilometres that's a lot of carbon. We could say then that we owe much of our coal to the invention of lignin.

Apart from the enormous amount of carbon burial going on, the multi-storey photosynthesis machine that was a Carboniferous forest was also producing an enormous amount of oxygen. Computer models indicate a rise of oxygen from 21% to 35% in the atmosphere due to the appearance of the Carboniferous forests. The high oxygen levels meant that wildfires were probably more common, being easily and often started by lightning strikes (Figure 2.3). The abundant oxygen also stimulated the metabolisms of insects and encouraged insects like dragon flies to grow very large (e.g. the Carboniferous fossil dragonfly *Meganeura* which had a wingspan of 60 cm).

What would the coal forests have looked like? Much of our detailed knowledge of the ecology of the swamp forests comes from the so-called *coal balls.* These unlikely objects are not coal at all but concretions of minerals that formed shortly after the burial of the plant material in the swamp (Figure 2.4). Two factors favour the lucky palaeontologist who finds a coal ball: one is that plant material (as well as insects and other flora and fauna) is often very well preserved inside; the other is that the material is not squashed as it would be in normal coal. The organic matter that accumulates to make coal is often compacted by a factor of 10 before it becomes coal. In a coal ball, the material is almost uncompressed, the hard minerals resisting most of the compaction. The first scientific description of coal balls was made in 1855 by Sir Joseph Dalton

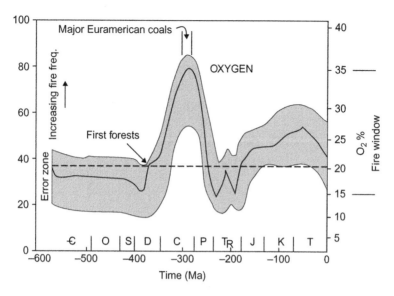

Figure 2.3 Modelled atmospheric levels of oxygen (%) (grey shows error zone) through geological time in millions of years (Ma). Also includes fire frequency, the 'fire window' and the major Carboniferous coal-forming period. From Scott (2000).

Figure 2.4 Coal balls from the United States. Department of Earth and Planetary Sciences, Washington University in St. Louis.

Hooker and Edward William Binney, who reported on examples in the coal seams of Yorkshire and Lancashire, England; but more recently much of the concentrated research on coal balls has switched to the United States. Over many years palaeontologists have thus been able to reconstruct with remarkable detail the conditions of the swamps.

It has long been known that lycopsids (club mosses) make up three quarters of the plant materials in coal balls, this suggests that the coal

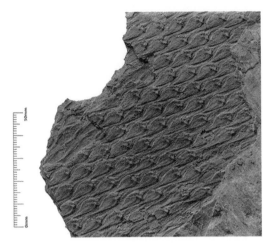

Figure 2.5 An example of Lepidodendron *from the collection of the British Geological Survey.* Lepidodendron glincanum *Eichwald, River Esk, Canonbie, BGS image P686121.* BGS copyright NERC.

swamps were dominated by a now-extinct primitive vascular tree known by the common name *scale tree* because of its bark which looks like fish or snake scales. Indeed in the nineteenth century the fossil trunks were often exhibited as giant snakes or lizards. Student geologists and palaeontologists know the distinctive diamond-shaped 'scales' as *Lepidodendron* (Figure 2.5).

The diamond-shaped repeated patterns in the photograph are leaf cushions. For most of its life the tree did not produce major branches but simply thrust up a crown of branches with narrow blade-like leaves. But the photosynthetic ability of these trees was not confined to the leaves. In fact the covering of the main trunk – the leaf cushions – was also capable of photosynthesis. We know this because stomata (the pores which allow CO_2 to diffuse into plants for photosynthesis) are present on the leaf cushions and the leaves. So the diamond-shaped patterns on the plant were not strictly bark in the sense that we know it. Probably the surface of the trunk was green – like leaves – and not grey or brown like the bark of more familiar modern trees.

This ability to photosynthesize over large areas meant that lycopsid trees like *Lepidodendron* grew fast and tall. It's estimated that *Lepidodendron* grew up to heights of 50 m with a trunk diameter of over 1 m and a rooting system spread over an area with diameter 24 m. These strange trees grew upward like poles at first until they made branches when reaching maturity. Some scientists have suggested that this would happen in as little as 10 years.

How did these primitive plants reach such great heights? Palaeontologists have studied the trunks in fossilized form, and structurally they can be considered as a set of concentric cylinders. The central one was probably rather soft but had a thin zone of strong lignin. Outside this was another soft zone surrounded itself by a thicker zone of lignin known as the periderm. This is very different from modern angiospermous or conifer trees which have a solid cylinder of lignin-reinforced tissue.

The roots of the *Lepidodendron* tree were huge and wide-spreading. These are well known to palaeontologists and fossil collectors and are common in Carboniferous rocks but oddly enough are known by another botanical name, *Stigmaria*. Sometimes *Stigmaria* is found attached to the lycopsid tree (Figure 2.6), but often is found separately, which is probably why different parts of the same plant have different palaeontological names.

The lycopsids of the Carboniferous coal swamps reproduced in a rather unusual way producing spores of two sizes, a microspore and a megaspore (called heterospory). These spores were produced in staggering numbers so as you might imagine the swamp was filled with spores. The study of spores and pollen is palynology, and a branch of palynology dealing with ancient spores and pollen (palaeopalynology) has grown up which studies these spores to understand the

Figure 2.6 A rarely seen upright lycopsid trunk with attached Stigmaria *root (visible just below the geological hammer). This specimen is from the Joggins Formation Cumberland Basin, Nova Scotia.* Courtesy M.C. Rygel.

environments in which coal was deposited, as well as producing a method of indirectly dating the coal.

Micro and megaspores of *Lepidodendron* and other similar lycopsid trees are very common in coal; in fact a single gram of coal can contain as many as a million *Lepidodendron* microspores, known as *Lycospora* by palynologists. *Lycospora* is very small about 30–40 μm (0.03 mm) in diameter and looks rather like a microscopic flying saucer with a narrow thin flange around the edge and a central disc-shaped 'fuselage' (Figure 2.7).

But what was the rest of the coal swamp like? For geologists to understand the wider character of the swamp we have to look beyond coal balls to sedimentary rocks that enclose the coal seams. In these, there are sedimentary structures and arrangements of sedimentary beds – as well as other fossils – which provide an insight.

Howard Falcon-Lang and colleagues have for many years explored and described the famous Joggins Formation of Nova Scotia, which contains numerous coal seams, but also some beautifully preserved fossils (including the remains of upright coal swamp trees with their roots

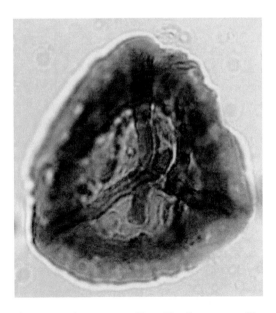

Figure 2.7 A specimen of Lycospora, *the microspore of lycopsid coal swamp trees. The spore is about 30 μm in diameter. BGS specimen from Waters Farm Borehole, Lancashire, England; Namurian, Sample MPA 18197, Specimen MPK 7635.*

in coal), as well as ancient river channels. From their detailed studies (Falcon-Lang et al., 2006) we have a picture of *Lepidodendron* peat-forming coastal forests, horsetail and seed-fern riparian (riverbank) forest all growing in semi-flooded conditions, with a rich collection of animals including molluscs, annelid worms, arthropods and tetrapods – including the earliest known reptiles. Inland of this were drier plains of cordaites forests (a primitive conifer-like plant) that were prone to wildfire. The coal swamps persisted for episodes of hundreds to thousands of years where peat accumulated at the same rate as water table rise so that the best conditions for preservation were created. Such an episode would create a single coal seam. Sometimes the sea rushed in to arrest coal development or to create coal with a slightly higher sulphur content (from the dissolved sulphates in the seawater).

The coal swamps stretched far to the east and west of Britain and Nova Scotia at the height of their development about 300 million years ago (Figure 1.5). Coal was being formed in what is now the northeastern and Midwest of the United States, the Maritimes of the US and Canada, North Africa, Britain, the North Sea – and east to eastern Europe and Asia.

HOW COAL FORMS

But plant material and peat are just the beginning of coal. Coal is essentially fossilized peat and peat forms in places where organic material collects. For peat to be preserved as coal it also has to be in a place where it won't be eroded. The best places have low elevation and are close to water and partially flooded but are also slowly subsiding. The subsidence means that if peat is growing it gets buried by a later layer of sediments and so is preserved. From this point onward the peat is 'embraced by the earth' in that it gets buried deeper and deeper. If the deep conditions are right, if temperature and pressure are not too low – and not too high – then coal will be formed.

Peat

Today peat covers 3% of the earth's surface and grows in many environments where accumulation of dead plant material exceeds the rate at which it is removed by atmospheric oxidation, burning or erosion. The unusual amount of preservation normally implies the presence of stagnant water which lowers the chance of fast rotting because oxygen is not present. Much of what we know about the environment of deposition of coal comes from study of modern peat, and there is a rather complex terminology attached to it. Rather than introducing many new words, an interested reader can probably get by with a few. A 'mire' is the general term for a place where peat of any type collects; a 'bog' is a mire which mainly gets its water from direct rainfall. Bogs are not always in lowland areas; they can form in hills and mountains. A 'swamp' is a mire normally at relatively low elevation which receives its water from the underlying water table. That these terms have rather restricted meanings reflects how important the source of water is in generating different kinds of peat and therefore different kinds of coal.

The environment in which peat forms also evolves through time. Swamps often start by infilling a lake or other poorly drained lowland hollow (Figure 2.8). Mud and plant debris slowly fills it up. The forest that the swamp supports may develop a great range of water-loving plants because of a variety of nutrients from groundwater. Because the swamp is low lying, rivers sometimes flood it with the result that river silt covers the floor of the swamp interrupting the pure supply of organic matter. If the swamp is close to sea level, the sea can cause a similar interruption. These influxes of sediment-laden water from out-side the swamp are responsible for the layers of shale or sandstone that break the pure coal up and introduce impurities.

In climates where rainfall is high throughout the year plants grow at great rates. The water table remains high and stagnation allows organic material to rot only very slowly so the peat becomes thicker and may begin to form a very low dome because year after year more plant debris is added. So the swamp begins to rise above the surround-ing land slightly. In this way the peat-forming process becomes more pure because it's more difficult for sediment from outside to contami-nate the peat. The swamp may also become a bog because the primary source of its water may be rainwater if the water table does not reach high enough to slow the rotting of the vegetation. It's for this reason that many coal geologists believe that raised swamps and bogs provide the highest quality coal. Sand or silt in the coal leads to less efficient combustion in power stations and to unburned ashy residue in the power station boiler.

But the formation of peat is only the first stage in a process that will eventually serve up coal. The peat has to be preserved in the long

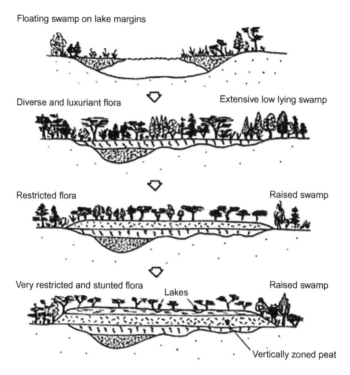

Floating swamp on lake margins

Diverse and luxuriant flora   Extensive low lying swamp

Restricted flora   Raised swamp

Very restricted and stunted flora   Lakes   Raised swamp

Vertically zoned peat

Figure 2.8 Evolution of a swamp. From McCabe (1987).

term – in other words over a geological timescale. A catastrophic weather event, for example a hurricane, may be enough to erode a raised swamp in a few days in which case the plant debris that has collected would be scattered by the sea. What are the best geological conditions to preserve peat and make coal? First the peat has got to be buried, then it's got to be heated and pressurized.

Subsidence

Burial happens easiest in subsiding land. Thick widespread deposits of peat are preserved in geological troughs called basins that subside over geological timescales. The whole of their land surface is slowly sinking. But nature must strike a very delicate balance in that the subsidence rate has to be tuned to the rate of production of the peat. If the rate of subsidence is too slow then the depressions in the landscape fill too quickly with sediment and not enough water is present to slow down the rotting and degradation of plant material. In this case peat will not form for a long period. On the other hand if subsidence is too fast, for example in a coastal area, the sea can inundate the swamp permanently — unlike in a short lived storm — and kill off the peat-forming plants and trees. If the rate of subsidence is right then peat can form for a long period, and a final thick inundation with new sediments caps and seals the peat and the peat begins its burial process. More and more layers of other sediments cover the layer and 'coalification' begins. This process essentially turns an unconsolidated 'mush' into a solid concentrated mass of organic matter with more and more concentrated carbon. Depending on how long and how warm the conditions are deep underground, the peat will transform into a series of coal types including brown coal or lignite early on — and eventually to bituminous coal and to anthracite. If the coal is heated too much or folded and faulted in mountain building, it can be burned away or become too broken up to be mineable.

Coalification

In coalification the plant debris is geochemically altered by heat and pressure over a very long period of geological time. Heat is the most important variable, and if it is available in a basin (e.g. from nearby volcanic activity) relatively young coal can be generated. But in most cases the temperature increase that applies is the mean geothermal gradient of the earth's crust which is about $25°C/1000\,m$ of descent. Most bituminous coal forms at temperatures of $100-150°C$. The quality or rank of the coal increases with depth. In the bituminous coal stage the organic material is heated to a point where hydrogen-rich compounds generate jelly-like bitumen which fills pore spaces in the coal. At this point the coal becomes denser and less porous. Further heating cracks the bitumen down to smaller molecules like carbon dioxide and methane. Methane adsorbs onto the surface of organic matter or exists as a free gas in pores in the coal. It's known as coal bed methane and is a considerable resource in some parts of the world.

Why was subsidence so common in the Carboniferous of Britain, eastern United States and eastern Europe? The map (Figure 2.9) shows coal swamps in a wide swath across the ancient continent of Euramerica. Years of geological study using palaeontological and palaeomagnetic methods have established the positions of the ancient continents very accurately, and it can be shown that a period of mountain building known as the Varsican Orogeny was partly responsible for the shape and structure of Euramerica, but more crucially it produced the conditions for the coal swamps to thrive.

Nowadays the remains of this mountain building can be seen in the hard metamorphic rock of the southwest of England and northern France, the Appalachian Mountains of northeastern United States and in the Hartz Mountains and Black Forest in Germany; but between 380 and 280 million years ago, the Varsican Orogeny created a range of very high mountains across northern France and Germany. What we see today in Cornwall is only the remains of those mountains which were similar in area and height to the modern Alps and were constructed, at their cores, of piles of folded and faulted rocks thrust together by converging plates.

But places where plates converge are complicated — not only does the inevitable crushing and crumpling happen, but also subsidence occurs over very wide areas. At the near edge of the mountains, subsidence can be very intense. To understand why this happens you have to imagine the sheer bulk and weight of the nearby mountain range sinking downward and pulling the earth's crust with it. It's a bit like the way a diving board dips down when a diver stands on the end. This area of strongly subsiding land in front of the mountains is known as a foreland basin. It's usually narrow and long.

Beyond the Varsican foreland basin, and further from the mountain front, ancient faults were reactivated and rather old crust (known as cratonic crust) began to subside as well. The subsidence in these two areas created the conditions for the coal swamps to be preserved for posterity in the form of saucer-shaped (or synclinal) heaps of sediments containing numerous coal seams. Where coal seams were preserved in the foreland basins, closer to the mountain front for example the

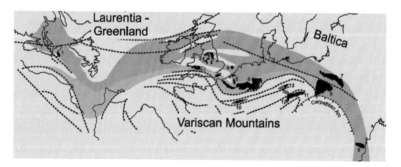

Figure 2.9 A reconstruction of the Variscan Mountains at about 300 million years ago. The dashed lines show the positions and trends of the main folds and faults. Light green — foreland basin; grey-green — cratonic basin; black — major coalfields. From Cleal et al. (2010).

South Wales coalfield (Figure 2.10), the sediments were preserved in steeper depressions, so that the coals seams are sometimes vertical (and therefore difficult to mine). The great depth at which the coal seams were buried in the centre of these depressions – as well as the proximity of mountain building – meant that some coal was converted to the highest rank (anthracite), as in South Wales.

The subsiding coal depressions of the cratonic area to the north in ancient Britain were more placid. Layers of sedimentary rocks and coal seams are mostly almost horizontal and rarely reached the rank of anthracite.

THE NEGATIVE GREENHOUSE

I mentioned earlier that a large coal-forming Westphalian lycopsid tree could have contained as much as 3200 kg of carbon, and if this was buried in a subsiding basin and converted into coal then most of the carbon was buried too. Cleal and Thomas (2005) estimated that coal swamps grew from an area of 500,000 km^2 around 310 million years ago to over 2 million square kilometres 10 million years later. Using a series of calculations assuming a certain lifespan for plants and density per square kilometre, they estimated that a stand of lycopsid trees might have captured and buried (through geological sedimentation and coalification)

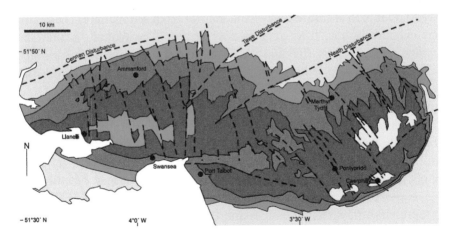

Figure 2.10 The South Wales coalfield formed in the foreland basin of the Varsican Orogeny. The grey colour is the South Wales Coal Measures and it is part of group of sedimentary rock layers in a deep saucer shape elongated west to east. The orange, blue and yellow are younger sedimentary layers that lie above the coal. From Cleal et al. (2010).

108–390 tonnes of carbon per hectare per year. Linking this figure with the maximum extent of the coal swamps gives a figure for the carbon capture rate of the whole coal swamps of almost 100 billion tonnes of carbon per year which would translate into a reduction of 44 ppmv of CO_2 per year in the Carboniferous atmosphere. Cleal and Thomas (2005) are the first to admit that these figures are rather approximate but when we compare the Carboniferous atmosphere to our own where human activities *add* a few parts CO_2 ppmv per decade it seems a very significant result.

Apart from the theory, is there any evidence that this huge level of carbon capture was happening in the Carboniferous? Earlier, I outlined how CO_2 has two main carbon isotope components, ^{12}C and ^{13}C, and their ratio is expressed as $\delta^{13}C$. The two isotopes are processed differently by plant photosynthesis. Plants prefer ^{12}C for their tissues but take in both isotopes, 'freezing' the ratio in their tissues. The $\delta^{13}C$ is therefore a record of the atmosphere at the time that the plants were growing. But also when plant growth is associated with large amounts of carbon burial – as in the Carboniferous coal swamps – the burial alters the balance of $\delta^{13}C$ in the atmosphere of the time. Lots of burial of carbon causes atmospheric $\delta^{13}C$ to increase and these increases are passed on to new plants. This means that $\delta^{13}C$ of plant tissue through time records increases in carbon burial or rapid expansion of swamp forests. Conversely a decline in the swamp forests would bring about a decrease in $\delta^{13}C$.

But fossil plant tissue $\delta^{13}C$ is rather difficult to measure. The easiest way to do it is to separate the organic matter from sedimentary rock samples chemically and then analyze the residue. The trouble is that the organic matter is very fine and it's difficult to see exactly what it's made of. If we take coal we know that it's likely to be made of carbon from lycopsid trees but also from other plants growing in the swamp. These might have slightly different $\delta^{13}C$. There might even be organic fragments from insects in the coal. What this means is that small differences in $\delta^{13}C$ in different samples might be due to something other than changes in the atmospheric $\delta^{13}C$ that the plants grew in – perhaps related to different plants or different kinds of organic matter. However larger variations in $\delta^{13}C$ captured in bulk organic matter from sedimentary rocks are believed to represent atmospheric changes and therefore changes related to large-scale burial of carbon. Large surveys of the $\delta^{13}C$ of bulk organic matter have been done and they tell a story of the evolving atmosphere and how the plants both modified it and were modified by it.

We can take the period of time between 450 and 250 million years ago (Figure 2.11), the period over which land plants appeared and began their colonization. The $\delta^{13}C$ of terrestrial organic matter (TOM) – in other words fragments of land plants in that period – varied between -26 and $-23‰$. The authors of the study (Strauss and Peters-Kottig, 2003) are circumspect about their data from fossil material older than 370 million years because of the uncertainty of the origin of some of the organic material they analyzed, however they recognize the rather large variations in $\delta^{13}C$ after 370 million years as very significant. The increase of $\delta^{13}C$ from -25.5 to $-23‰$ in the early stages of the Carboniferous about 360 million years ago was probably caused by the carbon burial associated with the expansion of land plants and the development of lignin. (Another measure of $\delta^{13}C$ from carbon in inorganic carbon compounds like limestone shows a similar change in $\delta^{13}C$ probably due to carbon burial in limestone in shallow seas at this time). The period between 330 and 300 million years ago over which the coal swamps grew is not associated with such an abrupt rise in $\delta^{13}C$. Strauss and Peters-Kottig believe the change is not so pronounced because of the effect of high oxygen levels in the atmosphere at this time.

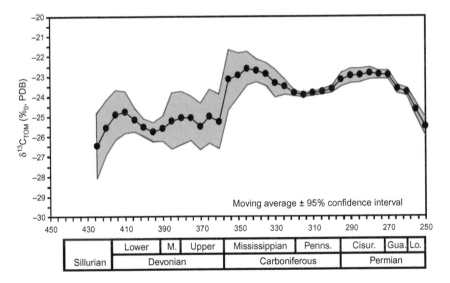

Figure 2.11 $\delta^{13}C$ of bulk TOM between 450 and 250 million years ago. The green band shows the error margin. The increase of $\delta^{13}C$ from -25.5 to $-23‰$ in the early stages of the Carboniferous about 360 million years ago was probably caused by the carbon burial associated with the expansion of land plants and the development of lignin. From Peters-Kottig et al. (2006).

Carbon capture does not only occur through burial of plant material as coal. Another large-scale geological process that does the same thing is weathering of silicate minerals. Because these minerals are what make up much of the hard rock of the earth, and particularly mountain chains, the creation and thrusting up of mountain chains is often associated with carbon capture because so much rock is newly exposed to fast weathering. This would have happened as the Variscan mountain range was created, and weathering would have been particularly intense as it was born close to the position of the Carboniferous equator (Figure 2.9). This occurring at about the same time as the expansion of the coal swamps would have created a powerful 'sink' for atmospheric CO_2. Within the last decade a theory has emerged that carbon capture through coal formation and the simultaneous creation of the Variscan mountain range lead to a 'negative greenhouse' condition, in other words the reverse of what we see now with increasing levels of CO_2 in the atmosphere. Though the evidence of atmospheric change from $\delta^{13}C$ is not particularly clear, there is abundant evidence for a glacial period, in fact one of the longest and geographically widespread in geological time, sometimes known as the Permo-Carboniferous glaciation which reached its height at about 300 million years ago. It affected the continents clustered around the South Pole (known as the supercontinent 'Gondwana'). The evidence is in the form of features in sedimentary rocks. There are for example 300-million-year-old boulder clay deposits in India, southern Africa, South America, Australia and Antarctica and Arabia. In Oman, in southern Arabia, spectacular scratches show where a glacier once scraped over an ancient rock surface. Nearby are cliffs which show rounded stones (called 'dropstones') inside fine-grained glacial lake sediments that could have only have got there because they dropped out of icebergs floating on the lake (Figure 2.12).

Many scientists believe that the most recent glacial period which began around 35 million years ago was caused by declining CO_2 in the atmosphere. For example Deconto and Pollard (2003) showed that as CO_2 declined, causing a reverse greenhouse effect, the East Antarctic ice sheet enlarged from small isolated ice caps into a large ice cap. There is similar evidence that through the last part of the Carboniferous between 330 and 300 million years ago, small ice caps coalesced to form a large ice cap which affected the southern hemisphere for several million years.

In the last few pages we have been examining a part of the carbon cycle, and the relationship between the atmosphere, climate and fossil fuels like coal. This connection is not new: Robert Berner in a short article in the journal *Nature* in 2003 pointed out how the long-term carbon cycle was capable of creating more than just coal. Another important product is what petroleum geologists call 'source rock' – as you might imagine this is the rock that provides the organic matter that generates oil and gas. Its formation is not dissimilar to that of coal, in that it involves plants photosynthesizing to make carbon compounds to make tissues, and those tissues accumulating and then being preserved deep in rocks. It's just that in source rocks – unlike coal – the plants (or bacteria) are microscopic and lived in oceans, seas and lakes. Rather than being preserved in peat in subsiding swamps, the organic matter of source rocks was originally preserved and captured in oxygen-deficient mud in deep seas or lakes. After burial, the processes that turn the organic matter into oil and gas are not much different to those that make coal from a mush of land plant material. There is one difference though – mostly the oil and gas once it's created escapes from the mud and finds its way into sandstone and limestone in structures called traps which allow the petroleum industry to drill for them.

Figure 2.12 Evidence of the Permo-Carboniferous glacial period about 300 million years ago that affected the southern continents. Left – scratches on an ancient rock surface left by Permo-Carboniferous glaciers. Right – a 'dropstone' dropped from floating ice into glacial lake sediments. Both from the Oman Huqf area. Photo M. H. Stephenson.

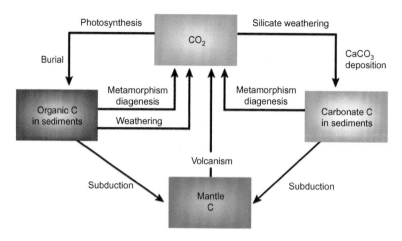

Figure 2.13 The long-term carbon cycle. On the left side photosynthesis creates organic matter in sediments (carbon burial or capture). This carbon can remain out of the atmosphere for a very long period in coal seams or source rock locked away in rocks for millions of years. But eventually the carbon can be returned through weathering of exposed coal or source rock. It can also – through subduction at plate tectonic boundaries – become part of the mantle. The mantle might return it by volcanoes. On the right we see silicate weathering (like the weathering of equatorial mountain ranges) leading to the mobilization of carbonate compounds and their eventual deposition as limestone. This is another way of capturing or burying carbon. From Berner (2003).

Looking at the carbon cycle diagram from Robert Berner's paper then, we need to realize that the box labelled 'Organic C in sediments' could be coal or the organic matter in source rock (Figure 2.13).

A final twist to this story is that glaciation itself has an influence on the creation of source rocks. It's not unusual in Earth history for concentrated source rock formation to take place after major glacial periods. Looking back in geological time, the next major glacial period before the 300-million-year-old Permo-Carboniferous event was the 'Hirnantian' glaciation which took place about 450 million years ago. Sedimentary rocks of this age show the same features – dropstones and glacial scratches – as those in Oman. Following the Hirnantian glacial period, ice melted and the sea level rose flooding glaciated lowland areas across what is now north Africa and the Middle East. The new shallow seas were fed by nutrients from the land and as the water warmed up they teemed with life. The hollows of the flooded landscape created low-oxygen havens for the preservation of organic matter. The result was a source rock of excellent quality that sits far below the surface across Libya, Algeria, Saudi Arabia and Oman. This source rock is known in Libya as the

Tannezuft Formation and it formed most of Libya's oil and gas. In Saudi Arabia the same rock is known as the Qusaiba Member and it is the source of most of that country's older oil and gas as well as the source of gas of the Qatari South Pars gas field which itself supplies much of Britain's liquefied natural gas (LNG). The US Geological Survey believes that the Qusaiba Member has produced 37 billion barrels of oil and 808 trillion cubic feet of gas that hasn't even been discovered yet!

A team from Leicester University (Page et al., 2007) are so convinced that the post-Hirnantian source rock buried carbon in huge amounts that they believe that its formation was responsible for cooling a climate that was rapidly warming out of control following the end of glaciation.

CAPTURING CARBON FAST ENOUGH

As we have seen, the creation of fossil fuels can in the long term cool climate, so why can't we encourage more natural carbon capture − planting more trees or seeding the oceans to make phytoplankton grow faster. Planting trees is one answer but it's been shown that modern forests are far less good at capturing carbon than the ancient Carboniferous variety, probably because modern ecosystems are much better at processing and metabolizing waste plant material so that it never gets buried. Whereas Carboniferous coal swamps could capture a hundred or more tonnes of carbon per hectare per year, a modern forest is capable of capturing less than 10. Seeding the phytoplankton of the oceans is seen by most scientists as very risky.

Our main problem is that we need to capture a lot of carbon very quickly. This is because the rate of increase of CO_2 in the atmosphere since the industrial revolution is the fastest that has ever been seen in history or in geological time. The only period that came near it was the Palaeocene-Eocene Thermal Maximum about 55 million years ago, and the evidence suggests that this caused widespread environmental chaos and extinction. In the future we face a stark choice − either we stop using fossil fuels like coal or we find a quick and effective way to capture and dispose of the CO_2. The rest of this book will look at this as an industrial process known as carbon capture and storage.

BIBLIOGRAPHY

Beerling, D.J., Berner, R.A., 2000. Impact of a Permo-Carboniferous high O_2 event on the terrestrial carbon cycle. Proc. Natl Acad. Sci. U.S.A. 97, 12428–12432.

Berner, R., 2003. The long-term carbon cycle, fossil fuels and atmospheric composition. Nature 426, 323–326.

Cleal, C.J., Opluštil, S., Thomas, B.A., Tenchov, Y., 2010. Late Moscovian terrestrial biotas and palaeoenvironments of Variscan Euramerica. Neth. J. Geosci. 88, 181–278.

Cleal, C.J., Thomas, B.A., 2005. Palaeozoic tropical rainforests and their effect on global climates: is the past the key to the present? Geobiology 3, 13–31.

Deconto, R., Pollard, D., 2003. A coupled climate–ice sheet modelling approach to the Early Cenozoic history of the Antarctic ice sheet. Palaeogeography, Palaeoclimatology, Palaeoecology 198, 39–52.

DiMichele, W.A., Phillips, T.L., 1985. Arborescent lycopod reproduction and paleoecology in a coal-swamp environment of late Middle Pennsylvanian age (Herrin Coal, Illinois, U.S.A.). Rev. Palaeobot. Palynol. 44, 1–26.

Falcon-Lang, H.J., Benton, M.J., Braddy, S.J., Davies, S.J., 2006. The Pennsylvanian tropical biome reconstructed from the Joggins Formation of Canada. J. Geol. Soc. London 163, 561–576.

McCabe, P.S., 1987. Facies studies of coal and cola-bearing strata: recent advances. Spec. Publ. Geol. Soc. 32, 51–66.

Page, A., et al., 2007. Were transgressive black shales a negative feedback modulating glacioeustasy in the Early Palaeozoic Icehouse? In: Deep Time Perspectives on Climate Change. The Geological Society, London, pp. 123–156.

Peters-Kottig, W., Strauss, H., Kerp, H., 2006. The land plant ^{13}C record and plant evolution in the Late Palaeozoic. Palaeogeogr. Palaeoclimatol. Palaeoecol. 240, 237–252.

Pfefferkorn, H.W., Gastaldo, R.A., DiMichele, W.A., Phillips, T.L. 2008. Pennsylvanian tropical floras from the United States as a record of changing climate. In; Fielding, C.R., Frank, T.D., Isbell, J.L. (Eds), Resolving the Late Paleozoic Ice Age in Time and Space. Geological Society of America Special Paper, 441, pp. 305–316.

Phillips, T.L., DiMichele, W.A., 1992. Comparative ecology and life-history biology of arborescent lycopsids in Late Carboniferous swamps of North America. Ann. Mo. Bot. Gard. 79, 560–588.

Scott, A.C., 2000. The pre-quaternary history of fire. Palaeogeogr. Palaeoclimatol. Palaeoecol. 164, 297–345.

Strauss, H., Peters-Kottig, W., 2003. The Paleozoic to Mesozoic carbon cycle revisited: the carbon isotopic composition of terrestrial organic matter. Geochem. Geophys. Geosyst. 4, 1–15.

Capturing Carbon Dioxide

There are 50,000 power stations in the world not all as large as Ratcliffe-on-Soar, and not all fuelled by coal. This may seem astonishing, but in an energy-hungry world electricity has to be generated on a huge scale. More and more electricity is being generated every year. In China for example an average-sized power station is opened every week. As we saw in the first chapter the developing world wants power for rural electrification and for new industry. But the developing world will also generate more electricity because this is the only way to decarbonize transport, in other words to make sure that our cars, trucks and buses don't produce emissions. The plan is to electrify transport — so a great expansion in electricity generation will be needed worldwide to charge up the cars — so power stations are here to stay.

But there's a hidden benefit in this concentration of emissions in power stations. If we can reduce their CO$_2$ outpourings we could make a big difference. Power stations are the biggest CO$_2$ producers and coal power stations are the worst amongst them, being the biggest and the dirtiest. But this also makes them the most attractive prospect for CO$_2$ capture because so much of the world's CO$_2$ emissions are concentrated there, and they don't move around, unlike cars. If we can capture their emissions we have a fighting chance of reducing CO$_2$ emissions substantially. In eastern England, for example, planners think that a single pipeline system serving the Humber area which has a lot of Britain's biggest power stations could capture 10% of Britain's emissions and dispose of them in the rocks under the North Sea. This could be operating as early as 2015. The good news is that CO$_2$ capture technology is known and well understood, though it will need to be used at a huge scale to make a difference.

Returning Carbon to Nature: Coal, Carbon Capture, and Storage. DOI: http://dx.doi.org/10.1016/B978-0-12-407671-6.00003-3

Keywords: *carbon capture; power stations; point sources; CO_2*

Ratcliffe-on-Soar is one of thousands of power stations in the world that burn fuel to make electricity, but in the process it produces around 10 million tonnes of CO_2 a year. A power station in the jargon of greenhouse gas science is an *industrial point source*, and these don't only include power stations — there are cement and ammonia factories, oil refineries and petrochemical works that also produce a lot of CO_2. A global survey conducted by the International Energy Agency showed that out of more than 14,000 point sources, the 7500 largest accounted for 99% of total global point source CO_2 emissions. Most of these are power stations. The statistics are staggering for single power stations for example Belchatow in Poland, presently the largest coal-fired power plant in the world, pumps about 34 million tonnes of CO_2 into the atmosphere annually.

How much space does all this CO_2 occupy? Going back to school chemistry you can calculate it using Avagadro's Hypothesis. One mole of a gas occupies 22.4 l at standard temperature and pressure (STP) which is 0°C and one atmosphere of air pressure (imagine a freezing day at sea level). One mole of CO_2 weighs 44 g. So 1 g of CO_2 will occupy 22.4/44 l at STP and 1 kg will occupy a space 1000 times larger, or 509 l. One tonne of CO_2 therefore has a volume of 509 cubic metres. A typical swimming pool is 25 m long and 15 m wide, so one tonne of CO_2 would fill it to a depth of 1.36 m (on a freezing day at sea level)! This is about the average depth of water in a swimming pool. But this is just one tonne. The Belchatow power plant emits over one tonne — or a swimming pool — of CO_2 every second!

The CO_2 has to be taken out of the atmosphere — that's clear — and in Chapter 2 I described the capturing power of natural organisms like land plants. Surely the simplest solution would be to plant more trees or prevent them being cut down? We know that modern forests are rather less good at permanently capturing CO_2 than say for example Carboniferous forests. But that doesn't mean we shouldn't try, after all forests are more than just places where CO_2 is absorbed, they are also a source of livelihood for forest dwellers and a haven of biodiversity.

There are attempts being made to prevent deforestation for example in the Brazilian state of Amazonas. Illegal logging has been a problem for years — at the current rate of deforestation, around one-third of the forest in Amazonas state will be gone by 2050, releasing a cumulative 3.5 billion tonnes of CO_2 into the atmosphere. REDD or 'reducing emissions from deforestation and degradation' a United Nations programme has a novel answer to this problem. In the Juma Sustainable Development Reserve in Amazonas, for example, local people will be paid to prevent trees from being cut down. Each family in the area has been given a debit card. If they prevent tree felling they get 50 reais (28 US dollars) a month credited to their accounts. This money comes from the rich world, where governments and companies that cannot reduce their own emissions cheaply are prepared to pay others to reduce emissions on their behalf (as 'carbon offsets'). Not cutting down trees in endangered areas prevent emissions that would otherwise have occurred, which gives untouched forest financial value − and provides people who live in the forest with an incentive to look after it.

But the problem we've created needs more accelerated action. There are good reasons to preserve the forests beyond CO_2 emissions, but the speed that it can work at sequestration is too slow to make a difference. For example although forests in the United States sequester around 10% of US CO_2 emissions, to reduce US CO_2 emissions by, say, 10%, would require the planting of an area bigger than the state of Texas with trees every 30 years. This is clearly not possible so we have to capture CO_2 at a much greater rate.

ARTIFICIAL CO_2 CAPTURE

Artificial capture of CO_2 isn't new at all. CO_2 has been extracted from the atmosphere in submarines for a long time. The air is made up of nitrogen (78%), oxygen (21%), argon (0.94%) and CO_2 (0.04%), but when we breathe it in our bodies consume the oxygen and convert it to CO_2. Our exhaled air contains about 4.5% carbon dioxide, a hundred times more concentrated than inhaled air. So in a confined space like a submarine CO_2 quickly accumulates and becomes poisonous. In a submarine CO_2 is removed chemically using soda lime (sodium hydroxide and calcium hydroxide). The carbon dioxide is trapped in the soda lime by a chemical reaction and removed from the air.

Why can't we use a similar reaction on a massive scale to capture CO_2 direct from the open atmosphere? The problem with this is the scale at which it would have to be done with such low concentrations of CO_2 and the huge volumes of air that would have to be processed. Actually, reactions like this are occurring all the time in areas of the world where aluminosilicate rocks like serpentinite are exposed at the surface. Serpentinite outcrops in Cyprus and Oman and the reactions that occur through natural weathering result in CO_2 being converted into solid carbonate minerals, but again not at sufficient rate to make a difference to our dilemma.

Apart from submarines there are industrial installations where CO_2 is separated from other gases at a relatively large scale, for example in the 'sweetening' of natural gas. Natural gas is a mixture of typically at least 90% methane, plus other hydrocarbons such as ethane and propane, as well as nitrogen, oxygen, carbon dioxide and sulphur compounds; and water. About half of global raw natural gas production contains CO_2 at concentrations averaging at least 4% by volume. Gas containing small volumes of these impurities can be used as fuel, but gas with higher amounts of CO_2 can't be burned safely. Also CO_2 and sulphur compounds dissolve in water to make acids which corrode pipes and machinery so it's good to get rid of them.

Gas sweetening means removing the acidic components like CO_2 and sulphur compounds and is a common process in refineries, petrochemical plants and natural gas processing plants. An example which has particular relevance to carbon capture and storage (CCS) is at Sleipner, a gas platform in the centre of the northern part of the North Sea. The field produces gas with unusually high levels (about 9%) of CO_2, but the Norwegian authorities want CO_2 levels less than 2.5% in the pipeline that goes to customers. In 1991 the Norwegian Government introduced a tax that penalized the release of CO_2 from gas sweetening. At the time of writing this is about 50 US dollars per tonne of CO_2 released. The company that operates the Sleipner Field – Statoil – obviously preferred not to release the gas and thereby avoid paying the tax, and so a special processing platform, Sleipner-T, was built to separate the CO_2 from the natural gas (Figure 3.1). Rather than pump the CO_2 separately ashore they investigated the idea of disposing of the gas in rocks about 1 km below the seabed but above the gas accumulation. They began this groundbreaking disposal in 1996 and are still injecting today so that more than 10 million tonnes of CO_2 are permanently trapped a kilometre below the seabed. This amount is about a year's worth of Radcliff-on-Soar's CO_2 emissions.

Figure 3.1 The Sleipner-T plant produces about 1 million tonnes of pure CO_2 per year and this is injected into a deep saline aquifer below the North Sea. Sleipner is the most important demonstration site for CCS technology in Europe. Photo: Statoil.

Table 3.1 CO_2 Concentration and Pressure in Waste Gases from Different Industrial Processes		
Source Type	CO_2 Concentration in Waste Gas (% by Volume)	Pressure of Gas Stream (kpa)
Coal-fired power station	12–14	100
Natural gas power station	7–10	100
Natural gas turbine	3–4	100
Iron blast furnace	Up to 27	200–300 before combustion; 100 after combustion
Cement kiln	14–33	100
Natural gas from a gas field	2–65	900–8000
Based on data from IPCC 2005: IPCC Special Report on Carbon Dioxide Capture and Storage. Prepared by Working Group III of the Intergovernmental Panel on Climate Change, Cambridge University Press.		

Sleipner is important because although it isn't related to power production, the scale at which CO_2 is being captured and injected is at least comparable to what might be needed in a modern power station. It's therefore a model for how a part of the industrial process of CCS might work in the future. We'll look at the geology of the CO_2 storage at Sleipner in Chapter 4.

So why concentrate on large point sources? As I mentioned earlier, the 7500 largest point sources on earth account for a high proportion of total CO_2 emissions. So if we turn our attention to these we have a chance of making a big difference. The concentration of CO_2 in the waste gas is also important (Table 3.1). This has to be high to allow chemical methods to work on an efficient scale. Large point sources sit in the landscape belching CO_2 at a high enough concentration to be treated. It makes sense to concentrate on these first, especially because they are often grouped together geographically. In the case of coal power stations the grouping is usually because there is an underlying coalfield (Figure 3.2). Many engineers and policy makers in CCS believe that large-scale CCS will first become economic and feasible in what are called 'clusters'. The Humber estuary in eastern England, Rotterdam and the Gippsland basin near Melbourne in Australia may well become CCS clusters in the near future.

How Do Power Stations Generate Electricity?

Coal power stations convert the heat energy of burnt coal into mechanical energy, which turns an electrical generator. Before the coal is burned, it's pulverized to the fineness of talcum powder, and then mixed with hot air and blown into the boiler. Burning in suspension in air, the coal provides the most complete combustion and maximum heat. Highly purified water, pumped through pipes inside the boiler, is turned into steam by the heat. The steam reaches high pressure and temperature and is piped to the turbine. The enormous pressure of the steam pushing against a series of giant turbine blades turns the turbine shaft and this is connected to the shaft of the generator, where magnets spin within wire coils to produce electricity. After turning the turbine, the steam is drawn into a condenser, a large chamber in the basement of the power station, where millions of litres of cool water from a nearby source (such as a river or lake) are pumped through a network of tubes running through the condenser. The cool water in the tubes converts the steam back into water that can be used over and over again in the plant Figure 3.3.

The important part of this process for our purposes is the exhaust gas produced by burning coal which is known as flue gas. It contains carbon dioxide (12–14%) and water vapour, as well as substances such as nitrogen oxides (NO_x – the 'x' can stand for 0, 1, 2 or 3), sulphur oxides (SO_x), mercury, traces of other metals, and fly ash, which is very fine solid silicon dioxide and calcium oxide.

Let's imagine what carbon capture would look like on my local power station Ratcliffe-on-Soar. In an early part of the book I described the flue gas desulphurization (FGD) equipment installed at the power station that extracts sulphur compounds from the flue gas.

Top 100 carbon-producing sites. Emissions in million tonnes CO₂

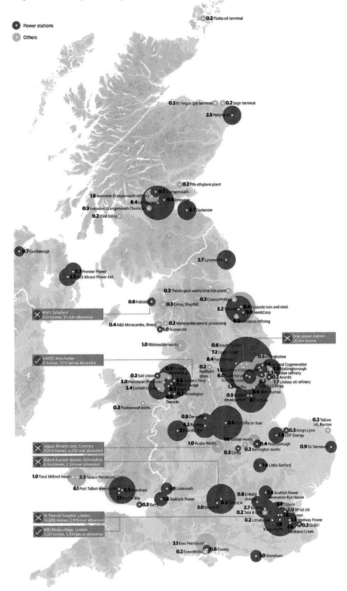

Figure 3.2 The top 100 CO₂ point sources in Britain. . Most of the point sources are power stations, and many are grouped together geographically, mainly over coal fields. It makes sense to target the groups of large emitters for CO₂ capture and to join them in pipelines to develop CCS clusters. The Humber estuary in eastern England could become a CCS cluster. From Kemp and Sola Kasim (2010).

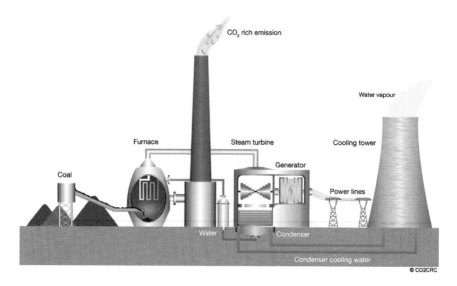

Figure 3.3 How a coal power station works. Copyright CO_2CRC.

The CO_2 capture equipment would be similar in that it would intercept CO_2 before it's released into the atmosphere – but the CO_2 capture equipment would be much larger than the FGD mainly because there is much more CO_2 in flue gas than sulphur compounds. In fact the CO_2 equipment if it was designed to capture all the power station's CO_2 would be enormous taking up an area of half a football pitch. The system would be so called 'post-combustion' because the CO_2 is removed after burning the coal (Figure 3.4). As this implies there is also a pre-combustion method as well. Most older power stations, if they are fitted with CO_2 capture in the future (known as 'retrofitted') would use post-combustion CO_2 removal.

There are several ways to capture the CO_2 after the coal is burnt. The favoured method at the moment is to bubble the flue gas through a solvent (likely to be a compound known as an amine) which absorbs the CO_2. The flue gas is pumped into the bottom of the absorber tower (Figure 3.5) while liquid solvent is poured in at the top. The buoyant gas passes up through the descending solvent and is absorbed by it so that mostly nitrogen is all that emerges from the top of the absorber tower. The solvent with CO_2 comes out of the bottom of the tower. For obvious reasons we can't afford to waste the loaded solvent and so this is pumped into a desorber. Here the solvent carrying its load of CO_2 is heated (by waste steam from the power station) to release the

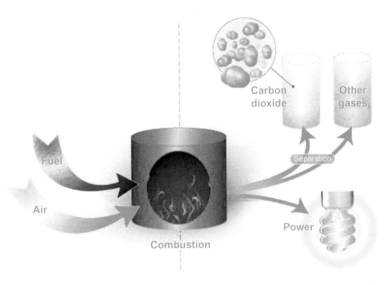

Figure 3.4 Post-combustion CO₂ removal. In this system CO₂ is removed from the flue gas after the coal is burnt. Copyright CO₂CRC.

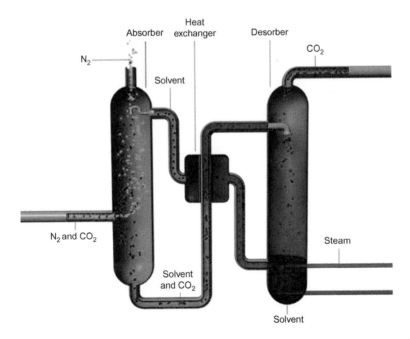

Figure 3.5 Absorption of CO₂ from flue gas. From RWE CCS website http://www.rwe.com/web/cms/en/2734/ rwe/innovation/projects-technologies/power-generation/fossil-fired-power-plants/CO₂-scrubbing/.

CO_2 which is collected at the top of the tower. The CO_2-free solvent is then returned to the absorber tower to be used again and the CO_2 is piped away. There are many methods designed to recycle heat as well, but even with these ways of saving energy in CO_2 removal, the process is energy intensive. Approximately 10−20% of the energy produced in the power station is used by the process if capture is working on all of the flue gas. So the bottom line is that CO_2 capture makes power stations up to 20% less efficient. But if the CO_2 that the station produces is disposed of permanently then the process renders the station's power producing processes almost carbon neutral, because more than 90% of CO_2 in flue gas can be captured.

What this means is that Ratcliffe-on-Soar power station − if 'retrofitted' with CO_2 capture − would be 20% less efficient than it is now, but would release almost no CO_2 into the atmosphere.

Although amine solvents can be recycled many times they become exhausted and would have to be disposed of. If very large amounts of capture was going on this might pose a problem. Using mechanical methods may provide a solution. Membrane separation, for example, is an established technology for separating nitrogen from air, but can be used to separate CO_2 from other gases. The membrane acts rather like a sieve after a pressure is applied, allowing CO_2 either to flow through or to remain, the result being that CO_2 is separated. In the adsorption method a solid substance can be designed to attract CO_2 either physically or chemically. The substance can then be treated to release the CO_2. In low temperature separation the flue gas can be cooled to a point where CO_2 condenses or freezes so that it can be removed.

What about pre-combustion capture? As its name implies, CO_2 is captured before the coal is burnt. This is because the coal is partially oxidized in air in a 'gasifier' to form a synthetic gas or 'syngas' before it goes anywhere near a power station boiler. The syngas has quite a high percentage of CO_2 but also has carbon monoxide (CO) and hydrogen. The CO_2 can be removed from the gas using absorption towers as in Figure 3.5, and reaction with water also allows conversion of CO to CO_2 so that this can be removed also. The power behind the syngas is in its hydrogen which can be used in a power station gas turbine to produce electricity.

If coal is burnt in an oxygen-rich atmosphere rather than air then much more CO_2 is produced in the flue gas, in fact most of the flue gas is CO_2. This is called oxyfuel combustion or oxyfiring. To create the oxygen-rich atmosphere, the nitrogen has to be removed (nitrogen is 78% by volume of air). This is done by cooling the air to $-180°$. At this temperature the oxygen in the air condenses as a liquid and is piped away to be used to burn the coal.

To give you an idea of the size and complexity of an oxyfuel plant, let's have a look at a place where a test plant is up and running – at the 1600 MW Schwarze Pumpe power station in eastern Germany where one of the less concentrated forms of coal called lignite is burnt to generate electricity. Here part of the power station has been converted to electricity generation and carbon capture using oxyfuel (Figure 3.6). This part will capture up to 100,000 tonnes of CO_2 a year while also generating 12 MW of electricity and 30 MW of heat, enough for about 1000 homes.

The oxyfuel system at Schwarze Pumpe is important because it is the first demonstration of oxyfuel as a carbon dioxide capture technology which allows the plant to go from lignite to electricity without releasing CO_2. In the jargon this is known as a 'full chain' demonstration (Figure 3.7).

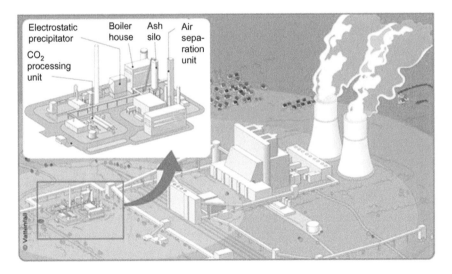

Figure 3.6 The oxyfuel carbon dioxide capture facility (in the blue rectangle) attached to the main power station at Schwarze Pumpe in eastern Germany. In the air separation unit, air is cooled to $-180°$ so that the oxygen in the air condenses as a liquid and is piped away to be used to burn the coal in the boiler house. In this oxygen-rich gas coal burns to produce flue gas which is almost entirely CO_2. Electricity plant processes based on oxyfuel combustion are sometimes referred to as 'zero emission', because the CO_2 stored is not a fraction or a proportion removed from the flue gas (as in the cases of pre- and post-combustion capture) but the entire flue gas stream itself. From the Vattenvall website.

Figure 3.7 The all-important CO_2 outflow pipes at the oxyfuel demonstration plant at Schwarze Pumpe power station. These pipes carry the gaseous CO_2 that has been separated during the oxyfuel process. When this photo was taken (June 2010) the plant was not operating hence the pipes are empty. Photographer: M.H. Stephenson.

TRANSPORTING CO₂

Where is the CO_2 from Schwarze Pumpe going I hear you say? Of course the capture of CO_2 is one part of the chain, but transport is another important part. I haven't introduced the intricacies of geological CO_2 disposal yet: but there is one simple fact to know — that not all rocks are suitable for CO_2 storage. Even if the rocks are right they may not be deep enough or they may be too deep. So options for storage are unlikely to exist right under the average power station. For example although there are rocks under Ratcliffe-on-Soar that might be suitable for CO_2 storage, they aren't deep enough. What this means is that captured CO_2 will in most cases need transporting. At Schwarze Pumpe, the CO_2 is likely to be disposed of in the rocks of the Altmark gas field in northern Germany, though at present the lack of German CCS legislation prevents CO_2 injection.

The transport of CO_2 by pipeline sounds rather odd, but like many industrial processes hidden from the public eye it is surprisingly common. Carbon dioxide pipelines are an established part of the pipeline infrastructure of the United States for example; in fact there are more than 3900 km of pipelines transporting 30 million tonnes of CO_2 annually, used mostly for injection into old oil fields to squeeze the last oil out. This is known as enhanced oil recovery, or EOR, and we'll look at this in more detail in Chapter 4. The carbon dioxide for EOR is mostly taken not only from natural underground reserves, but also from gas sweetening operations, ammonia factories and coal gasifiers.

Pipelines in the United States commonly transport carbon dioxide as a gas. A compressor pushes the gas through the pipeline and there are mini-compressors along the way to keep the gas moving. The CO_2 must be free of hydrogen sulphide and not contain liquid water or water vapour because this can corrode the carbon manganese steel pipelines. A stainless steel pipeline can resist corrosion but is very expensive.

I said earlier that Ratcliffe-on-Soar power station doesn't have suitable deep underground CO_2 storage and that transport would be needed if it were to be retrofitted with CO_2 capture. How would the transport arrangements for the power station look? Would they be feasible and economic? For this we need to think of CCS in geographical clusters, and the nearest future cluster to Ratcliffe-on-Soar might be in eastern England around the Humber River.

The Humber CCS Cluster

If you drive north through eastern England on one of the main highways there are three things that strike you immediately: flat prosperous agricultural land, rivers and power stations. No sooner has one risen on the horizon and then disappeared in your wing mirror its cooling towers pouring out white steam, than another appears. The Humber is England's engine room – a group of stations that take advantage of nearby coal, water and flat land to power much of Britain. The main coal-fired power stations are Drax, Eggborough and Ferrybridge C but there are also six gas power stations. There are 12 industrial sites or power stations in the area that produce more than 1 million tonnes of CO_2 per year, the largest of which is Drax, Britain's biggest power station, which is capable of producing 10% of England's electricity (Figure 3.8).

But the Humber area also produces 90 million tonnes of CO_2 emissions annually, about 60 from power stations and factories. This is

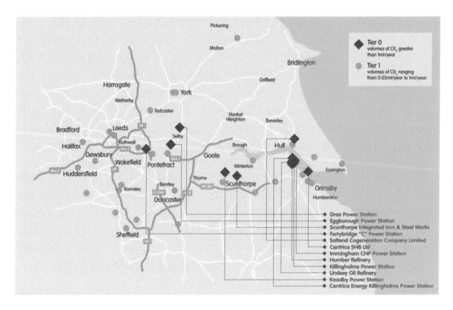

Figure 3.8 Point sources in the Humber area in eastern England. 'Tier 0' emitters are the largest with annual out-pourings of greater than 1 million tonnes of CO_2. 'Tier 1' point sources emit 50,000 to 1 million tonnes of CO_2 per year. From Yorkshire Forward: A carbon capture and storage network for Yorkshire and Humber.

quite a high proportion of Britain's total carbon dioxide emissions of about 450 million tonnes (2010 estimate). Engineers and planners say that if the main Humber point sources were fitted with CO_2 capture and linked up by a pipeline then 10% of the Britain's emissions could be disposed of. Is there geological storage space nearby? The answer is yes – and we'll examine this in Chapter 4 – but let's look in more detail at the proposed pipeline.

Piping CO_2 is of course not trivial. To keep costs down it makes sense to connect up enough emitters as possible to a single network of pipes and to connect this with a single disposal site. But for sites spread across the Humber, the problem is rather like the famous mathematical riddle of the Seven Bridges of Königsberg where the idea was to find a walk through the city that would cross each bridge once and only once. Nevertheless the proposers, a team at the Leeds Engineering company CO_2Sense, think they've come up with such a pipeline network, capable of carrying up to 60 million tonnes of CO_2 per year by 2030. They envisage two separate phases: in the first phase two to three point sources south of the Humber River will be connected. In the second phase three more point sources in the Aire

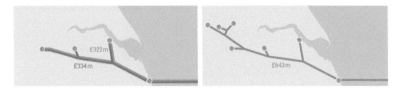

Figure 3.9 What a pipeline system for Humber CCS cluster might look like. Two separate phases are envisaged: in the first phase the pipeline will be connected to two or three emitters south of the Humber River (left). In the second phase CCS is developed by a further three emitters, in the Aire Valley (right). The costs of the different options are also shown. From A carbon capture and storage network for Yorkshire and the Humber: pre-front end engineering study executive summary, CO_2 Sense (2010).

Valley to the north will come on-stream. The pipeline leads to the coast where it launches offshore. The distance from the coast to the offshore storage facilities would be about 100 km (Figure 3.9).

Where does this leave the power stations along the River Trent 30 km southwest from the Humber: West Burton and Cottam — and what about more distant Ratcliffe-on-Soar about another 50 km south? Could these be connected to the Humber cluster? It all depends on the economics. Perhaps in the future these would add another 25 million tonnes of CO_2 per year to the pipeline.

Another reason why CO_2 transport is not trivial is a pipeline of this type will have to be built to the right size to allow later point sources to contribute to the stream. Different power stations also produce slightly different flue gases because they burn different kinds of coal and gas and the separated CO_2 will inevitably have small amounts of impurities which can have powerful effects on the way that CO_2 behaves in the pipeline, and possibly in the rock formation (more about this later).

So the pipeline has to be built to accommodate different gas types. It might be convenient in the future when huge amounts of CO_2 are being moved around the country to pipe it in a concentrated state known as 'supercritical'. Normally CO_2 is a gas and at higher temperature and pressure it's a liquid — but beyond certain limits of temperature and pressure (see Chapter 4) it becomes 'supercritical' in that it's dense like a liquid but flows very easily like a gas. It's hard to imagine exactly because it would be difficult to observe, but supercritical CO_2 would act a little like WD40, the lubricating oil. The advantage of piping CO_2 in this runny supercritical form would be that a lot of CO_2 could be moved in a small pipe — more than if it was in gas form — and

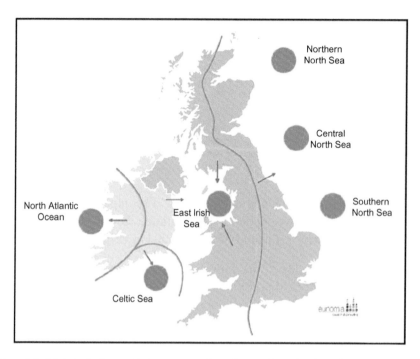

Figure 3.10 Britain in the future might have two main disposal sites and a CCS industry that is divided between west and east. From Eunomia (2011).

it wouldn't need much pumping because it flows so easily. Of course the pressurization and heating would cost a lot and a balance would have to be struck between costs of compression and savings gained by fast flowing dense CO_2.

In a future Britain where carbon dioxide is captured from every power station and even industrial point sources, there would have to be several clusters and they would depend on the location of the geological storage and the geography of the point sources. Some see a Britain with a CCS 'watershed' running up its spine where the point sources of the east of Britain pump their CO_2 to the east for disposal in the North Sea, while to the west the East Irish Sea might become a focus for geological disposal (Figure 3.10).

The East Irish Sea geological disposal areas are the gas fields of Morecambe Bay and Hamilton and these might receive CO_2 from point sources in southern Scotland, the northwest, North Wales, east Ireland and Northern Ireland (Figure 3.11).

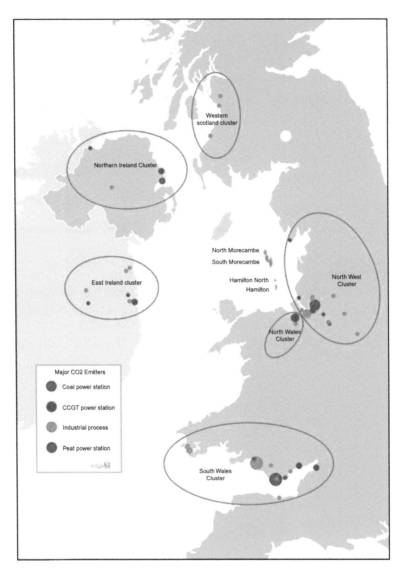

Figure 3.11 Point sources around the east Irish Sea. From Eunomia (2011).

Other regions may become clusters – the key being the proximity of large point sources and geological storage potential. One area where the idea of a cluster is very advanced is in south-eastern Australia around the Latrobe Valley east of Melbourne and offshore in the Bass Strait. The Latrobe Valley contains the world's second largest lignite deposit and produces more than 90% of the state of Victoria's electricity but also most of Victoria's total CO_2 emissions of around 122 million tonnes per year. Offshore in the Gippsland Basin geologists believe that there is space for 20 gigatonnes of CO_2 which represents hundreds of years of storage for the Latrobe Valley's CO_2 output.

BIBLIOGRAPHY

CarbonNet Project. <www.dpi.vic.gov.au/carbonnet>.

CO_2CRC. <http://www.CO_2crc.com.au/>.

CO_2Sense, 2010. A Carbon Capture and Storage network for Yorkshire and the Humber: Pre-Front End Engineering Study Executive summary.

DECC UK emissions statistics. <http://www.decc.gov.uk/en/content/cms/statistics/climate_stats/gg_emissions/uk_emissions/uk_emissions.aspx>.

Economist, September 24, 2009. A new climate treaty could provide a highly effective way to reduce carbon emissions by paying people to not cut down forests.

Eunomia, 2011. The East Irish Sea CCS Cluster: A Conceptual Design, Technical Report.

Kemp, A.G., Kasim, S.A., 2010. A futuristic least-cost optimisation model of CO_2 transportation and storage in the UK/UK continental shelf. Energy Policy 38, 3652–3667.

Metz, B., Davidson, O., de Coninck, H., Loos, M., Meyer, L. (Eds.), 2005. Carbon Dioxide Capture and Storage IPCC. Cambridge University Press, UK.

Yorkshire Forward. A carbon capture and storage network for Yorkshire and Humber.

CHAPTER 4

Returning Carbon to Nature

Ratcliffe-on-Soar power station might be connected into a carbon capture and storage (CCS) cluster in the next decade. We know it's possible to capture its CO_2 and to pipe it away because this has been done before and is being done now on quite a large scale elsewhere. But what about the place where we want to store it – the deep underground? Surely we can never see what this CO_2 would look like deep in rocks, surely we can never really know what will happen to the CO_2 – how it will move and how it will change? Well the answer is that we know a surprising amount. Startling pictures of the so-called 'Sleipner plume' actually show CO_2 collecting in sandstone rock 1 km below the seabed of the North Sea. Computer simulations show how CO_2 is likely to move after it's been injected and even how it might change over hundreds and thousands of years. The good news is that CO_2 is likely to dissolve in water far below the surface and probably also react to form solid carbonate minerals in spaces in the rock. So geologists are confident that the longer the CO_2 is down there, the less likely it will be able to escape.

Returning Carbon to Nature: Coal, Carbon Capture, and Storage. DOI: http://dx.doi.org/10.1016/B978-0-12-407671-6.00004-5

The final ironical twist to the story is that it's even possible to store CO_2 in tiny cracks and spaces in coal. In South Africa geologists are considering this very option following the carbon cycle from the formation of coal to its combustion and release of carbon into the atmosphere as CO_2, and then back as injected CO_2 into the very rock that it came from.

Keywords: *carbon storage; saline aquifer; storage space; CO_2; depleted field*

Almost 5 km below the woods, lakes and rivers of Jackson, Mississippi, a hundred miles north of New Orleans is 200 million tonnes of natural CO_2. The CO_2 is trapped under a huge arch-shaped structure known as the Pisgah anticline and appears not to have moved for a very long time. The CO_2 is not leaking to the surface and due to the huge pressure and high temperature at this great depth is in the high density runny supercritical form. The accumulation was discovered around 1960 by the oil company Shell who were drilling for oil and gas, and at first no one knew what to do with all the CO_2. The Mississippi oil and gas company Denbury Resources quickly came upon a solution though – to use the CO_2 in enhanced oil recovery (EOR). Denbury Resources bought the rights to drill and tap into the CO_2 accumulation and piped the CO_2 to old depleting oilfields in Mississippi where it was used to squeeze out the remaining oil. In the Little Creek oilfield oil production was tripled from 1300 to 3500 barrels per day. In the nearby Mallalieu field similar gains in oil production were achieved by EOR using CO_2.

But this isn't the prime reason why geologists are interested in the Pisgah anticline and its huge reserve of CO_2. It's because the CO_2 has been there for 70 million years without moving, and because the size of accumulation is what we're aiming at in industrial CO_2 disposal. In other words carbon capture and storage (CCS) scientists want to study the Pisgah anticline because this is what an industrial CO_2 disposal site might look like in the future.

How is the CO_2 held in the rocks deep below the Mississippi countryside and why doesn't it flow up to the surface? A lot of the CO_2 is contained within sandstones that were formed in a desert during the Jurassic period about 150 million years ago. This doesn't mean that the CO_2 is 150 million years old, it's actually younger – but it's 'soaked' into this rock in supercritical form. A typical sandstone rock is exactly what it says in the name – sand turned into stone. The sand particles are still there – broadly spherical in shape – but they aren't loose but are bound together usually by some other mineral that came along later and cemented the sand to make it solid. But it's not the particles that matter in a sandstone for an oil geologist or a CCS scientist – it's the gaps between the particles. Though the particles are cemented together, there are always spaces left in between known as pores. One way of picturing this is to imagine a box of oranges. The oranges are the sand particles on a very large scale and the spaces between the oranges are the pores. It isn't hard to see that these pores are likely to be connected. If you were to pour water into the box, probably a lot of water would go in, depending on how tightly packed the oranges are. There might even be an equal amount of water and 'orange' in the box. In this case the pores are 50% of the volume of the box. Geologists would say that the 'porosity' is 50%.

Now the porosity of the sandstone in the main CO_2-bearing parts of the Pisgah anticline is between 8% and 15%. This is the volume that is occupied by CO_2. This doesn't sound a lot but when you realize how huge a volume of rock there is under the Pisgah anticline – the structure is 30 km long and 8 km wide – this adds up to a lot of space. That's why there's estimated to be 200 million tonnes down there (Figure 4.1).

What about the second question – why doesn't the CO_2 flow to the earth's surface? This is because there are rock layers above the sandstone that are so compact and dense that they don't allow gases or liquids to pass through them. Remember the box of oranges – well in this case the oranges are replaced by *much* smaller and flatter objects – something like thousands of coins or counters laid flat inside the box. This is the three-dimensional structure of a rock like shale – very different to sandstone. The result is that porosity is lower. More importantly the pores or the spaces between the particles are less well connected. In the box of oranges a liquid or gas can easily pass through because the spaces are large and because they are connected. In a denser more compact rock like shale the spaces are small and poorly connected. A geologist would say that shale has low permeability, whereas sandstone has a high permeability. Some rocks like rock salt have no pores because they are crystalline all the way through, in other words completely solid, and so have no permeability at all.

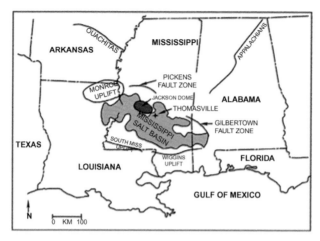

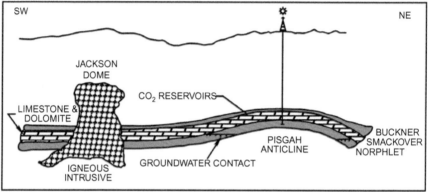

Figure 4.1 The Pisgah anticline in Mississippi, a deep natural arch shape in which CO_2 has collected and from which CO_2 is now extracted for EOR. The rocks under the arch contain 200 million tonnes of natural CO_2 which has been there since it was formed around 70 million years ago. From Studlick et al. (1990).

The reason for the CO_2 remaining in the Pisgah sandstone is that above it there are layers of very low permeability salt-rich rocks and above those is also a layer of shale about 30 m thick. A layer like the shale or the salt-rich rock is known as a caprock or a seal in geology, because it stops gases or liquids moving upward. In fact, without seals and caprocks deep underground there would be no oil or gas fields because oil and gas are also stopped from moving upward and so collect in useful accumulations. One way of thinking about a seal is to imagine the damp course in the wall of a house – usually made of a waterproof rubbery layer that you can put in between layers of bricks. The damp course stops moisture moving up through the wall as a seal or caprock stops CO_2, gas or oil moving upward (Figure 4.2).

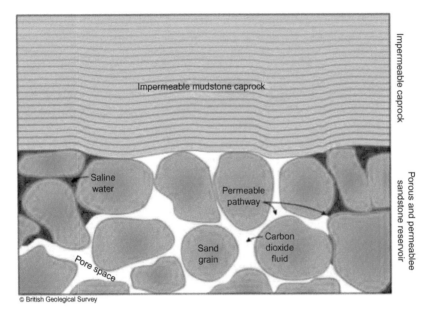

Impermeable caprock

Porous and permeable sandstone reservoir

© British Geological Survey

Figure 4.2 A representation of the difference between a porous and permeable rock and a non-porous, imperme-able rock (seal or caprock). The spaces between the sand grains in sandstone can contain CO_2 as in the Pisgah anticline or saline water. In a hydrocarbon field the spaces can contain oil or gas. Courtesy Michelle Bentham. BGS copyright NERC.

Where do very large amounts of natural underground CO_2 come from? Deep volcanic activity can create CO_2 just as volcanoes at the sur-face can. Also chemical reactions affecting rocks, particularly limestone, can make CO_2. We are quite fortunate that at Pisgah we know precisely where the CO_2 came from. The most starling fact about the CO_2 in the Pisgah anticline is that it's very old, about 70 million years old. The gas is 99% CO_2 but has small amounts of methane, nitrogen and H_2S and also very small amounts of other rarer gases. You might remember from Chapters 1 and 2 that the carbon in CO_2 is made up of two iso-topes mainly. These are ^{13}C and ^{12}C and the ratio between them can be expressed as $\delta^{13}C$. The $\delta^{13}C$ value of the CO_2 in the Pisgah anticline ranges from -3.55 to -2.57‰ which is much higher than would be expected from living sources or from once-living material such as coal or limestone. This suggests a non-living source for the CO_2. The gas also contains isotopic ratios of rare gases such as helium and argon that suggest that the CO_2 came from very deep in the earth, ultimately from the mantle. Nearby is a large mass of igneous rock – the Jackson Dome (Figure 4.1). One of the useful things about igneous rocks – crystalline rocks that are essentially frozen magma — is that they contain ratios of radioactive elements themselves which allow us to date them quite accurately. By chance the Jackson Dome rock is known to be 70 million years old — in other words it froze from molten magma deep underground about 70 million years ago. Its formation probably produced the CO_2 in the nearby Pisgah anticline. This is very useful to CCS geologists because it suggests that the CO_2 trapped under the seal hasn't moved for 70 million years — because it's *still there*.

Let's consider this for a moment. Geologists are used to thinking about very long periods of time and are fond of making analogies. The dinosaurs disappeared about 65 million years ago; the Panama isthmus closed about 4 million years ago at about the same time that *Homo sapiens* appeared; the ice age in Europe ended about 10,000 years ago. But the Pisgah anticline CO_2 has been trapped under a seal for longer than any of these. This gives geologists a lot of optimism in CCS as a way to combat climate change because if we can store CO_2 for 70 million years that's far longer than our present climate crisis is likely to last. More of this later.

But of course artificially creating underground stores of CO_2 like the one at Pisgah anticline will not be easy.

CREATING AN UNDERGROUND CO_2 STORE

For a start to put CO_2 into the pores of rocks we'll have to inject it. This means we'll have to force whatever is already in the pores out of the way.

I know that I'm going to get in trouble with my fellow geologists for showing you this picture — a magnificent mountainside in western Iran (Figure 4.3).

I've chosen the picture because it shows a large anticline — the kind of natural arch in the rock layers that I mentioned earlier in the chapter. The picture is a very useful way to illustrate how a geological structure could be used to store CO_2. The problem of using this picture to illustrate CO_2 storage is first that you can see the structure.

We will never be able to actually see a structure that we chose to inject CO_2 into. We'll know all about them — their shape and size but

Figure 4.3 In the foreground is a large anticline — the kind of natural arch in rock layers that might be used to store CO₂. From http://www.fas.org/irp/imint/docs/rst/Sect2/Sect2_6.html credit to FAS.org.

we won't be able to see them. The structures we chose will have to be deep — at least 800–1000 m down. This is another reason why the picture isn't quite right — because the structure you can see is too shallow. I'll explain why later. But the photograph is a good one because you can see what some of the deep structures we might use would look like.

Now if you also imagine that the structure is composed partly of caprock or seal and partly of sandstone then the anticline in the picture might be a storage structure. For the purposes of this demonstration, the green arch-shaped layer is a caprock and the yellow layer underneath is sandstone. A new word I'll introduce here is 'reservoir' for the sandstone because the rock is capable of storing a fluid in its pores. So the green is caprock and the yellow is reservoir (Figure 4.4, left).

If you were to drill a hole through the structure and pump CO₂ down into the reservoir you would first of all have to push what's in the pores of the reservoir out of the way. This is something that I didn't mention earlier in the chapter — the fact that most pores below the water table to a fairly great depth are filled with water. Water at shallow depths is commonly fresh — and is tapped by water wells the world over. Deeper water is usually very saline, and at the depths that

Figure 4.4 Filling the structure with CO_2. The green is caprock and the yellow is reservoir. In the picture on the right, CO_2 has been injected and is confined under the structure. Credit to FAS.org.

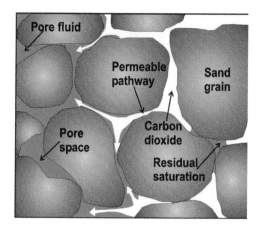

Figure 4.5 Pumping CO_2 into a reservoir will push the pore fluid (usually saline water) out of the way. Some of the saline water might remain clinging to the sand grains – this is known as residual saturation. Courtesy Michelle Bentham. BGS copyright NERC.

CCS is planned the pore water would be far too saline to be of use. Nevertheless this saline water has to be pushed out of the way so that the CO_2 can occupy the pores. The arrows in the right hand picture in Figure 4.4 show this happening. Figure 4.5 shows this happening at a microscopic scale.

Of course the saline water has to have somewhere to go. This is not simple either. If there's somewhere the displaced water can go then plenty of CO_2 can be injected into the rock. If the saline water can't move for any reason, then you have a problem because there will be a rise in the pressure of the fluids (the CO_2 and the saline water) in the reservoir.

Figure 4.6 A road cutting in the United States exposing a cliff face. There are two faults and some imaginary seal rocks shown in green on the photo on the right. Courtesy Sam Holloway and Gary Kirby. BGS copyright NERC.

Figure 4.6 illustrates this. It's a photo of a road cutting in the United States exposing a cliff face (Figure 4.6, left). The picture is useful because we can see two small geological faults (fractures along which some movement has occurred) shown in green on the photo on the right.

I've also for the purposes of this demonstration shown two caprock or seal layers (the green bands). Sometimes faults act as seals as well preventing fluids from moving across them. So we have a parallelogram-shaped structure in the photo outlined by seals above and below and to the side, or perhaps a three-dimensional parallelogram if we imagine it far below the earth's surface.

What would happen if we inject CO_2 into this structure? Well the CO_2 would push some of the saline pore water out of the way but not for very long because water is not compressible. Very quickly the CO_2 and the water would become highly pressurised if the CO_2 injection pumps were powerful enough – rather like a pressure cooker. The pressure could become big enough to fracture the seal above the reservoir which wouldn't be good for confining the CO_2. So *reservoir pressure* is important in CCS.

It's also important to realize that the pressure in the fluid (the CO_2 or the saline water) is separate from the pressure of the rock. In many ways the pressure of the fluid is independent of the rock around it. This may sound rather counter-intuitive but the pressure of the pore water is often equal simply to the weight of the water above it – in other words all the water up to its upper surface at the water table.

But the rock that surrounds it is often supporting a much greater pressure — that of the weight of the *rock* — perhaps several kilometres worth of it. This weight is being supported by the particles of the rock — all the sand grains — rather like the columns of a cathedral supporting its roof. Usually the pore water pressure is much lower than the rock pressure — think of the low pressure air between the columns in a cathedral — but if it's increased to the same as the rock pressure the rocks above can fracture.

In Figure 4.4 I showed CO_2 collecting immediately under the arch of the anticline. There is a reason for this. Though the CO_2 would be pushing the pore water aside, the CO_2 fluid (remember it would be in super-critical form below 800 m depth) would also be less dense than the pore water and so would be buoyant. So it will tend to collect under the highest part of the structure. If you continued to inject CO_2 it would fill the arch more and more. Below it would be ordinary pore water.

This is the principle of the 'trap' — that CO_2 is trapped just like in the Pisgah anticline — for a very long period underneath an impermeable arch. CCS geologists call this kind of trap a 'physical trap', and the process 'physical trapping'. This is because simple physical processes — buoyancy, flow and pressure — govern it. The same principle applies to oil and gas fields — buoyant hydrocarbons also sit under natural impermeable arches all over the world. The fact that physical trapping can be seen to have worked in the Pisgah anticline and many other natural CO_2 fields — as well as in oil and gas fields — gives geologists a lot of confidence that artificial CO_2 stores can be created underground.

I've come a long way into this book without explaining the phases of CO_2 and this rather alarming term 'supercritical'. The fact that CO_2 can exist in a dense form isn't just useful for piping CO_2 around. It's also useful for disposing of it underground because if we inject CO_2 in supercritical form it will take up less space. This means a lot of CO_2 can be disposed of in relatively little pore space. We'll have a look at how much will be needed by a typical power station in the next section, but first let's look at CO_2's phase behaviour.

The graph (Figure 4.7) shows these phases. We're all familiar with dry ice — the solid form of CO_2 — and of course we're continuously breathing in the gaseous form, but supercritical is the form it has beyond the critical point (31.1°C and 73.8 bar). Why is this important for storage?

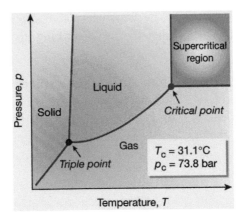

Figure 4.7 Phase diagram of CO₂. From the following article: Green chemistry: Designed to dissolve, Walter Leitner, Nature 405, 129–130 (11 May 2000) doi:10.1038/35012181.

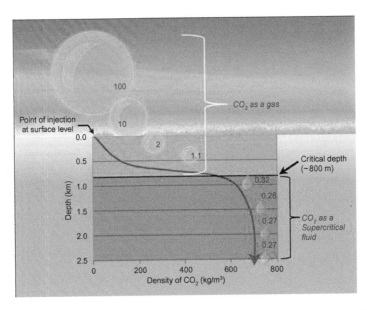

Figure 4.8 If you start with a CO₂ volume of an arbitrary 100 units by the time you get down to a few kilometres below the surface, the CO₂ is only taking up a quarter of one unit — in other words the CO₂ is taking up only a quarter of 1% of its original volume at the surface. From the NETL website: http://www.netl.doe.gov/technologies/carbon_seq/FAQs/carboncapture3.html.

The reason is simple: the deeper you go the higher the pore fluid pressure. Typically if you inject below about 800 m depth the pore fluid pressure will be so great that it will exceed the critical point and the CO_2 will become supercritical. This is shown in Figure 4.8. If you start with a CO_2 volume of an arbitrary 100 units by the time you get down to a few kilometres below the surface the CO_2 is only taking up a quarter of one unit — in other words it is taking up only a quarter of 1% of its original volume at the surface.

You might remember from an earlier chapter that Belchatow power station produces a swimming pool worth of CO_2 every second at standard temperature and pressure. In supercritical form that same CO_2 would occupy only about one cubic metre. So a year of Belchatow's emissions would occupy a pore space of less than 130,000 cubic metres.

But now I hear you ask — how much pore space is there underground and can it realistic take the emissions that we want to dispose of? Let's have a look at one particular sandstone layer known by geologists as the Bunter sandstone.

The picture (Figure 4.9) shows the Bunter sandstone in a quarry near Scunthorpe in eastern England, and the inset picture shows what the same sandstone looks like under a microscope. You can see the yellowish sand particles and even in this photo it's clear that the

Figure 4.9 The Bunter sandstone in a quarry near Scunthorpe in eastern England; the inset picture shows what the sandstone under a microscope. Courtesy Sam Holloway and Graham Lott. BGS copyright NERC.

Bunter sandstone has pores between the sand particles — in fact the porosity of the sandstone is 15–20% and the permeability is high. We know quite a lot about this sandstone because it forms outcrops in the midlands of England and on the coast of the south of England. In Nottingham the sandstone is well known because it has caves which were at one time occupied by people, and there is a still a pub in the city which is partly built into the sandstone.

Geologists have had their eye on the Bunter sandstone for a long while, not just because a pub was built into it but also because it's a promising rock to store CO_2. The aim is not to store CO_2 in the Bunter sandstone on land though, but to look at its potential offshore in the North Sea, not far from the Humber estuary (Figure 4.10).

The map (Figure 4.10) shows the extent of the Bunter sandstone layer both on land and under the seabed. The layer underlies much of England in a Y-shaped pattern but only outcrops in thin strips around northeastern England and the Midlands. It plunges deep below the surface in eastern England and continues offshore. In places it's more

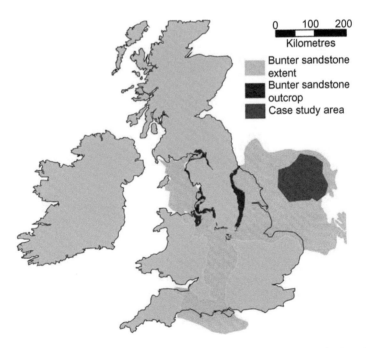

Figure 4.10 The outcrop and underground extent of the Bunter sandstone. Much of it is under the seabed of the North Sea. From Smith et al. (2011).

than 3 km below ground level. We know where it is, how thick it is and quite a lot about its porosity and permeability from seismic surveys (more about these later) and deep boreholes onshore and offshore that have penetrated the layer. In fact our knowledge is so good that we've been able to build three-dimensional computer models of the layer in certain areas. One such area is shown on the map (Figure 4.10) in red, a case study area for CO_2 storage.

Now three-dimensional models don't look very good on two-dimensional pages but the picture below shows what the case study area looks like (Figure 4.11).

The Bunter sandstone is 200–350 m thick but over the huge area of the case study this is pretty thin – which is why the layer appears so thin in the diagram. The layer is also very uneven – it looks like a floor mat that has been rumpled – and the different colours and contours show the depths of the layer below the seabed. The red colour means shallow – this is where the layer comes within a few tens of metres of the seabed or even 'outcrops' on the seabed. The blue means where the layer is deep, in fact several kilometres below the seabed.

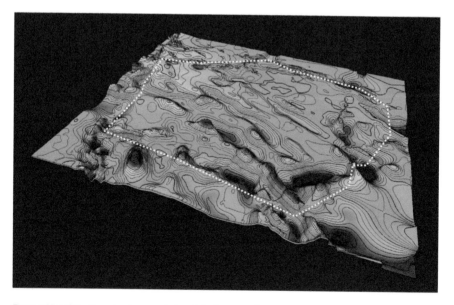

Figure 4.11 A three-dimensional representation of the Bunter sandstone in part of the North Sea. The white dotted line shows the edge of the 'case study' area shown in Figure 4.10. The area within the white dotted line is about 10,000 km². Courtesy Andy Chadwick. BGS copyright NERC.

The layer is like this because it's been folded over very long periods of time and also because it's been distorted by ancient salt movements.

But enough of this. How can we use this to demonstrate CO_2 storage? The answer is we can simulate storage using our knowledge of the layer. In this process we imagine injection of CO_2 and build a computer model to show us what will happen. Modelling or simulation like this is a very established method in modern science and engineering – in fact very little is built or made without making a computer model of it first – to try things out. Here we try CCS out on an offshore layer to see how much CO_2 it will take. The model will obviously have drawbacks which we'll discuss later, but it's a useful guide.

So what did the modellers find? In their simulation they imagined 12 injection points injecting a combined rate of 33 million tonnes of CO_2 per year for 50 years (Figure 4.12). In real life this would look like 12 oil rigs out on the North Sea injecting CO_2 rather than taking oil out. Thirty-three million tonnes of CO_2 per year is the equivalent of three large power stations.

What you see in the map are the 12 points coloured on a grey background. The colours represent CO_2 saturation – in other words the

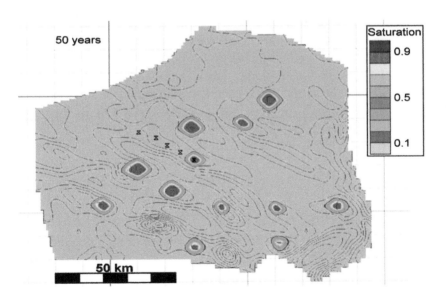

Figure 4.12 The result of the simulation in the case study area: injecting 33 million tonnes of CO_2 per year for 50 years. From Smith et al. (2011).

proportion of the pores that is occupied by CO_2, red representing the highest saturations. The 12 points show where the accumulated CO_2 over 50 years (in fact 1650 million tonnes) has gone. The answer is that it's stayed very close to the injection wells, which is what you would expect. The injection sites are below arches or anticlines so that the CO_2 will be trapped. What's very interesting is that the CO_2 after the simulation occupies *less than 1%* of the total pore volume of the case study area. The bottom line is that the rock has easily accommodated 50 years emissions from three large power stations and there is awful lot more space left.

As you can imagine this kind of simulation makes geologists quite excited about the possibilities for CCS. But there are problems, not least the fact that the geology in the models is grossly simplified. What I mean by this is that lots of assumptions have been made in the creation of the computer model – one simple one is that we assume that the sandstone layer is the same all the way through – that the rock doesn't vary. Well if you sit in the pub in Nottingham with its walls of Bunter sandstone and look at the hollowed out structure you can see that the rock does vary – in colour, in structure and probably in porosity.

Another problem is pressure. When you inject CO_2 you have to push the pore water out of the way. If there's somewhere for the water to go it's alright. If there isn't then the pore fluid pressure in the reservoir will go up. To illustrate earlier this I showed a photo (Figure 4.6) of a road cutting with faults and seals. What I was illustrating there was what CCS geologists would call a 'closed system' – completely sealed. An ideal situation for injection would be an open system (but with a seal or caprock above) – but what is the situation in the Bunter sandstone out there under the North Sea?

The answer is that we don't know for certain because the area is so large – so the modellers decided to simulate closed and open systems to see what would happen during injection. This is the beauty of simulation – that you can try out lots of scenarios.

Let's look at the 'open' system result for the computer model first (Figure 4.13).

A little extra explanation is needed here for you to understand this result. First the pore fluid pressure is shown in colour. Red areas

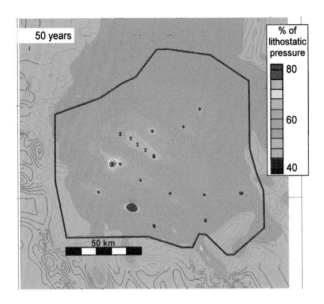

Figure 4.13 Result of the simulation of an open system for the Bunter sandstone layer, again after 50 years of injection. This picture doesn't show the CO₂ injected, but the pore fluid pressure in the rock. From Smith et al. (2011).

indicate high pore fluid pressure and green and blue low pressure – the red dots are the twelve injection points. The good news is that most of the map is green or blue – which is what you might expect because this is an open system and the pore water can move, relieving pressure.

However all is not simple. There are red areas in the map also. What do they mean?

To understand this we have to know the scale for pressure that the modellers used. Rather than show an absolute value for the pressure the modellers used a neat trick – they expressed the pressure as a percentage of lithostatic pressure. Lithostatic pressure is the technical term for the pressure caused by the weight of rock above the layer – which is considerable if the layer is deep. In the example of the road cutting (Figure 4.7) the lithostatic pressure is what 'keeps the lid' on the reservoir or the caprock stopping it from fracturing. As I said in that example, if the pore fluid pressure is close to the lithostatic pressure then the seal could be broken and the CO₂ could leak out.

So the percentage of lithostatic pressure is important – and you can see that the percentage is quite high in a few places, even though this is an open system.

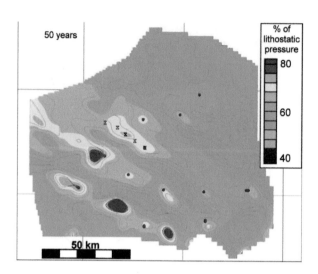

Figure 4.14 Result of the simulation of a closed system for the Bunter sandstone layer, again after 50 years of injection. This picture doesn't show the CO₂ injected, but the pore fluid pressure in the rock. From Smith et al. (2011).

Let's look at the result of the simulation assuming a closed system (Figure 4.14).

As you might expect, the overall pore fluid pressure in the reservoir is higher (there is more red and green than blue) because this is a closed system simulation. The red parts are also larger and more numerous.

In fact the modellers and geologists think that this part of the Bunter sandstone is probably 'partially open' in that there are structures around the edge of the modelled area that probably form barriers to pore fluid flow, so the situation after 50 years of imaginary injection is that a lot of CO_2 can be accommodated, that most of the reservoir remains only lightly pressured – but that a few points reach reservoir pressures that are quite high in relation to lithostatic pressure. A quick look at the maps will show that these points are also places where the reservoir is shallow. The reason for this is simple – in these places the lithostatic pressure is low (because the rocks are not deep) and so similar fluid pressure across the reservoir reaches a higher proportion of lithostatic pressure in the shallow parts.

Now the observant reader here will have immediately thought that if pore water is being moved out of the way it has to go somewhere. In a very large reservoir without barriers to flow, pore water might be moved but not very far. As I said before, the Bunter sandstone actually reaches the seabed in a couple of places. If you are injecting CO_2 somewhere in the reservoir – it could be tens of kilometres away – a very small amount of pore water is likely to be pushed out through the seabed. In other words the sandstone would leak pore water into the seawater above it. This would generally not matter because the salinity of the pore water in such shallow rocks is likely to be the same as the sea itself. But the trick is obviously not to let out any CO_2, after all what's the point in letting it out if you've gone to that trouble of injecting it?

So the key is to inject in structures away from shallow parts of the reservoir, and make sure that the places you inject into have capacities that won't allow any spills into neighbouring structures. So you'll notice that most of the simulation injection points in the maps (Figures 4.13 and 4.14) are in arch-shaped structures, but that these are deep and not close to shallow structures that if injected with CO_2 might leak into the sea. Remember also that we would not inject into shallow structures anyway because we want CO_2 to be in supercritical form. Some geologists rather like the idea of small amounts of pore water being pushed into the sea because this makes more room for CO_2. Some even suggest that holes could be drilled into reservoirs and water deliberately extracted – to relieve pore fluid pressure to increase space for CO_2.

We've looked at a reservoir over a wide area; now let's look at an individual 'trap' using simulations like the one I described above. What we'd like to know is how buoyant supercritical CO_2 behaves after it's been injected into the reservoir.

In Figure 4.15 the modellers have taken a single structure of the Bunter sandstone and simulated 50 years of CO_2 injection. The simulation can be viewed as a short film, but here I've shown five 'stills' from the film showing the CO_2 at different periods of time after the start of injection: at 3, 10, 50, 150 and 1781 years.

Remember that the CO_2 is buoyant and will therefore seek the highest place, trying to keep above the saline water pore fluid that also occupies the reservoir. In this simulation, injection only takes place for 50 years and after that the injection is stopped and we observe what happens to the CO_2. You might ask why we're so interested in the CO_2 so far ahead – but this is something we've really got to think about.

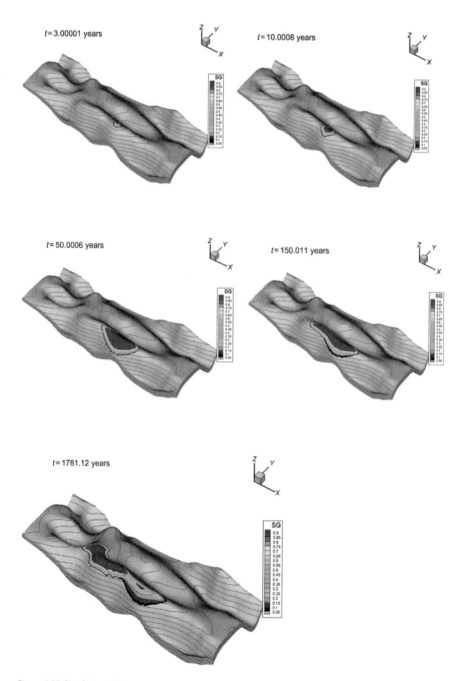

Figure 4.15 Simulation of buoyant trapping of CO₂. Courtesy of David Noy and Andy Chadwick. BGS copyright NERC.

In the first picture ($t = 3$ years), injection has just started and so the saturated area of CO_2 is very small. The highest saturation is close to the injection well which is what we would expect. In the second picture (10 years) you can see what 10 years of accumulated CO_2 looks like — more or less a circular spot in plan view. The main process affecting the shape of the patch of CO_2 appears to be the radial force of injection. At 50 years you can see the process of buoyancy beginning to act in that the CO_2 is beginning to move up the arch-shaped structure. At 150 years (100 years after the *end* of injection), the CO_2 has moved a lot — into the highest parts of the structure. After 1781 years, it has almost left the original arch structure in its quest for the highest parts of the reservoir. And there it should stay under a seal or caprock — perhaps for 70 million years as in the Pisgah anticline.

You might ask why the modellers didn't continue the simulation for thousands of years. The answer is that the physical assumptions of the model are probably not valid over a very long period because the CO_2 starts to dissolve in the pore water. This is the next very interesting thing about CO_2 disposal — and a new complication — that CO_2 doesn't just sit there in the pores — it actually dissolves and can also *react* with the rocks themselves. We'll look at these complications later in the chapter.

READY-MADE STORAGE

Much of this chapter has been about how geologists are trying to make a CO_2 store. In fact lots of ready-made stores already exist — these are oil and gas fields that are empty or nearly empty.

Earlier I described how oil and gas collect naturally in traps under seals or caprocks because they have lower density than the surrounding pore water. Gas has a lower density than oil and so will often occupy the highest part of the arch (Figure 4.16).

In Figure 4.16, Well 1 would produce oil and gas and Well 2 would produce water. But through time Well 1 would empty much of the oil and gas out of the arch and might begin to produce only water. In the oil industry this is known as a 'depleted' or 'depleting field'. It's known for certain that many of these traps have held oil and gas for tens of millions of years (like the Pisgah anticline has held CO_2) and so many

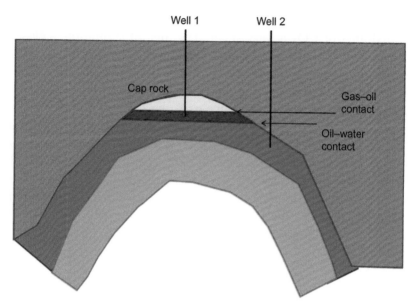

Figure 4.16 Oil and gas under a caprock arch.

geologists think they are prime candidates for CO_2 disposal once the oil and gas has been removed. The trap is like a tank or container that you can use for whatever gas you want to store. All over the world natural gas is stored in such traps. The Rough Field, for example, 18 miles off the east coast of Yorkshire, eastern England is used to store up to 2.8 billion cubic metres of natural gas (most of Britain's stored gas). The gas originally came from Norway or Qatar in the Middle East but needs to be stored long term for periods when gas becomes scarce, for example in winter. But the Rough Field used to be a natural gas field until the mid 1980s when it ran out of gas. So natural gas can be pumped in or out of a reservoir.

If CO_2 could also be pumped in and out of the reservoir this leads to an obvious question. Will we ever re-use the CO_2? At the moment it's hard to know whether large amounts of CO_2 could be useful. This is why CCS geologists use the word 'storage' rather than 'disposal' in the phrase 'carbon capture and storage' — because they're hedging their bets!

Depleted oil and gas fields are 'ready-made' in the sense that there are wells, platforms and pipes (from the time when oil and gas were extracted) that could be used for CO_2. Their structure, shape, porosity and permeability are very well known because oil companies spend enormous amounts of money drilling into these structures (a single off-shore oil well can cost tens of millions of dollars), as well as making computer simulations of the way the oil or gas flows. The simulations can be adapted for CO_2. Oil companies have long experience of 'managing' reservoirs because they are their main asset. So the companies take very good care of the reservoir and the trap. One example is the management of the caprock. As oil and gas is pumped from a reservoir this can cause it to shrink slightly – often because it becomes more compressible. The key is to make sure that any shrinkage is too small to have an effect on the caprock. If the caprock was to subside slightly down into the reservoir, it might crack. This is something that the oil company would want to avoid – they don't want the caprock leaking because they will lose some of their oil and gas. This type of very careful management of the reservoir and the caprock (the trap) is what will be needed in the future for CCS.

In the last few years the British Government has been trying to encourage the development of CCS by offering a financial incentive to encourage companies in CCS. As part of the 'Carbon Capture and Storage (CCS) Competition' four industrial consortia have been short listed to negotiate with Government for financial help of £ 1 billion to part-fund an industrial scale project. All of the four projects are likely to use structures in the southern North Sea for disposal offshore, possibly in the Bunter sandstone (the contents of the bids are secret). Some of the bids mention that depleted oil and gas fields are the main targets and at least one bid is likely to link up with the Humber cluster, which I mentioned in the last chapter. It's possible therefore that perhaps later in this decade power stations like Ratcliffe-on-Soar, and those along the Humber River will be connected as a *full chain* from power generation, to capture to storage in the North Sea. To companies, depleted oil and gas fields offer the lowest risk and cheapest options for CCS – being ready-made they are the 'low hanging fruit' of storage structures. Their main drawback - which we'll look at later – is that their capacities tend to be rather small. While they offer a short term solution in the early days of CCS, we need much bigger storage as well.

BIGGER STORAGE

In the last chapter I mentioned the longest-running industrial scale CO_2 storage project in the world at Sleipner. It's also unique because its main aim is CO_2 emissions reduction rather than, for example, EOR.

As it happens, at Sleipner CO_2 is not being injected into a depleted oil or gas field but into a thick layer of sandstone without a strong arch structure but with an absolutely enormous CO_2 capacity. The drilling platform stands in 80 m of water and the CO_2 injection point is at a depth of about 1012 m below sea level in a sandstone layer called the Utsira Sand. The injection point is about 200 m below the top of the Utsira Sand.

Figure 4.17 is a cross section of the rock layers below the seabed near Sleipner, running east to west through the Utsira Sand. It's been constructed from sesimic surveys carried out by oil companies looking for oil and gas in the North Sea. The yellow vertical lines show the position of oil and gas wells. The Utsira Sand is shown in yellow and the lower seal or caprock in green.

Just a few words about how a cross section like this is constructed. The cross section is really a plot of sound waves which are generated on a boat or a vehicle (if on land). The boat (in the case of the cross section in Figure 4.17) generates a very loud sound and then records the echoes of that sound bouncing off the boundaries between rock layers far below the seabed. Analysing how quickly and how strongly the waves are reflected back indicates the arrangement of rock layers. The vertical scale in Figure 4.17 is not in metres but in 'two way travel time' because it still shows how long the echoes took to leave the ship and reflect back to it. But in this case the units still approximate to metres.

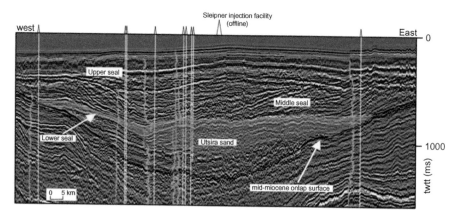

Figure 4.17 The rock layers below the seabed near Sleipner. From Chadwick et al. (2008) BGS copyright NERC. Reproduced with permission of the C02STORE Consortium.

The first thing you notice about the diagram is the Utsira Sand doesn't form an arch shape. In fact its shape is rather indistinct. The sandstone doesn't look like a layer at all. This is partly because the cross section is very highly vertically exaggerated – in other words squashed sideways – so the layer looks thick and narrow. In fact the Utsira Sand extends rather widely under the North Sea. Figure 4.18 shows that the sandstone extends 50 km south of Sleipner, 50 km east and west of Sleipner and hundreds of kilometres north into the north Atlantic. It varies from about 300 m thick (the darker areas) to very thin. The Sliepner area is where it's thickest. So the large area and great thickness means that the Utsira Sand has an enormous volume. This coupled with the fact that the sandstone has a porosity in the range of 27–42% and high permeability means that it is could have a very large capacity for CO_2.

Above the Utsira Sand in the Sleipner area is a low-permeability mudstone 50–100 m thick which is known as the 'primary seal' because it's the main barrier between the CO_2 that's been injected and

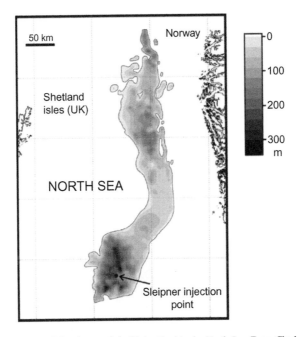

Figure 4.18 The thickness and distribution of the Utsira Sand in the North Sea. From Chadwick et al. (2008). BGS copyright NERC. Reproduced with permission of the C02STORE Consortium.

the seabed (Figure 4.17). Above this is another seal again of mudstone; and above that an upper seal of ancient glacial sediments.

What does the CO_2 that's been injected look like? You probably wouldn't expect to be able to see it, but in fact through some very clever analysis of seismic 'sound' waves – the same ones that allow us to make a cross section – it's possible to build up an image of the CO_2. In fact boundaries between patches of supercritical CO_2 and pore water also reflect seismic waves and these can be plotted like the boundaries between rock layers. Because Statoil, the company that operates the Sleipner site were so interested in the progress of their CO_2 injection, they did several seismic surveys. The idea was to build up a picture of the CO_2 as it collected. They did seismic surveys in 1994 (before any CO_2 was injected) and then in 2001, 2004, 2006 and 2008. Injection started in 1996.

The images they made are shown in Figure 4.19. It shows the extraordinary growth of the mass of CO_2 – known by geologists as 'the plume'. The picture on the far left shows the full thickness of the Utsira Sand with its upper and lower surfaces. In the image from 2001, you can see a series of red, grey and blue stripes. These indicate 'layers' of supercritical CO_2 'soaked' into the rock. The layers are getting wider and probably thicker right up until 2008 and until today, because injection is still going on.

Remember that the injection point is 200 m below the top of the layer, so not far from its base. Because the supercritical CO_2 has a lower density than the saline pore water it moves upward as far as the base of the seal and collects under it, rather like water under a damp course in a wall. Another extraordinary finding of the seismic survey was that it took about a year for the first CO_2 – the upper edge of the plume – to reach the base of the seal.

Figure 4.19 The growth of the CO_2 plume at Sleipner. Courtesy Andy Chadwick. BGS copyright NERC.

The final set of diagrams (Figure 4.20) I want to show you needs a bit of extra explanation.

There are three mottled blue rectangles. Each rectangle is not a cross section but a plan view. The black shape in the same position in each rectangle is the outline of the plume – also in plan view. It's what the plume looks like from above, rather than from the side as in Figure 4.19. So the three blue rectangles are like maps of the same level within the seal above the Utsira Sand. But there's one complication: rather than show physical features, the maps show differences in the reflection data between different seismic surveys. The first shows differences between reflections that were received in 1994 from those received in 2001 – so it shows differences in the character of the seal rock between those dates. You can see there are no data differences that correspond to the shape of the plume in any of the three maps. The logic of the geologists who made the maps is that *if there are no differences in the seal directly above the plume then it very unlikely that any CO_2 is leaking into the seal.* This is a very important finding because it supports the idea that the CO_2 will stay underneath the seal.

But what about the lack of an obvious arch or other trap shape? In the case of Sleipner and the Utsira Sand the shape doesn't matter

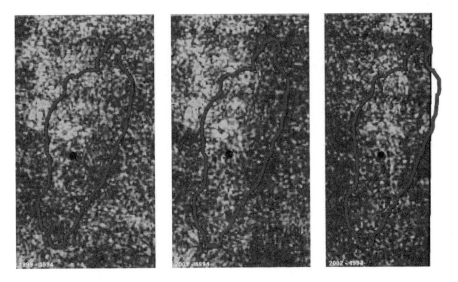

Figure 4.20 The difference between the earliest seismic reflection data and the later data for a level within the seal above the Utsira Sand. From Chadwick et al. (2008). BGS copyright NERC. Reproduced with permission of the C02STORE Consortium.

because the layer itself is so large and because the seal covers it comprehensively. Big deep layers that are full of saline water like this are very common and are known by CCS geologists as 'deep saline aquifers' or 'deep saline formations'. They are very attractive as places to store or dispose of CO_2 because they offer such large capacities — much more than do depleted oil and gas fields. In a deep saline aquifer like the Utsira Sand if the CO_2 moves sideways the seal will always be there to confine it. It won't move very far because the movement is so slow. Even if it does move the plume will leave part of itself behind in the pores. If you look back at Figure 4.5 you'll see how small 'bubbles' of water are left behind in pores as CO_2 surges through the reservoir. This can also happen to supercritical CO_2 — in other words as the plume moves through the reservoir — rising because of its buoyancy — it leaves 'bubbles' of CO_2 behind it. These bubbles are fixed in the rock as a residue. Think of when you wring a towel out to remove the water — there will also be a residual amount left. If a plume in a very large deep saline aquifer moved a large distance it would leave much of its CO_2 behind it rather like the vapour trail of an airplane across the sky.

This way of trapping the CO_2 in the rock is known by geologists as 'hydrodynamic trapping' and it doesn't need an arch shape, or a seal. If the deep saline aquifer was thick enough and you injected very deep into the layer the CO_2 would be retained within the layer as a residue and may never reach the top of the layer.

There are other processes we also need to consider over very long periods because this is when CO_2 begins to behave rather differently to oil and gas.

LONG-TERM BEHAVIOUR OF CO_2

The key difference between CO_2 and say natural gas is that CO_2 is *reactive*. It's capable of reacting with pore water and with rock particles, depending on their composition.

Some CO_2 will eventually dissolve in the pore water. Like hydrodynamic trapping this will keep the CO_2 in the reservoir. The reason is that once CO_2 is dissolved, it doesn't exist as a separate chemical phase because it's part of the pore water. This means there is no longer any buoyant supercritical CO_2 and therefore no force to drive the CO_2 up through the reservoir. The CO_2 is locked in the water and will not go upward. This is therefore another kind of trap, known as 'solubility trapping'.

Geologists would like to maximize the amount of solubility trapping in the reservoir because it does away with the need for a physical seal or arch structure. But there is a problem in that CO_2 is only slightly soluble in water. Engineers will probably try to manage injection so that solubility trapping starts as soon as possible. The solubility of CO_2 in water increases with increasing pressure, but decreases with increasing temperature and salinity (Figure 4.21).

The CO_2 will also dissolve rapidly at first but if the pore water isn't replenished it will become saturated with CO_2 and won't be able to dissolve any more. If there is some natural flow that brings fresh pore water in contact with CO_2 then dissolution will continue.

Of course anyone that has done some chemistry at school will know that when CO_2 dissolves in water it makes a weak acid. So there will be a lot of acidic pore water around that will react much more with minerals in the particles of the reservoir rock than ordinary pore water. What's good about this is that it's yet another method of trapping

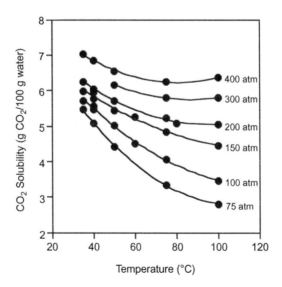

Figure 4.21 Solubility of CO_2 in pure water with changing pressure and temperature. From Stephenson and Rochelle (2010), http://www.rsc.org/Education/EiC/Restricted/2010/July/HaveYourCoalAndBurnIt.asp – Reproduced by permission of The Royal Society of Chemistry.

CO_2 because once the CO_2 has become part of the pore water it can be further fixed in the rock if it reacts with the rock to make a new solid mineral. New minerals will be formed which will sit in the pore spaces. Then the CO_2 is really fixed and won't be going anywhere for a long time. This is called 'mineral trapping'. Relatively fast reactions – perhaps over days or months – might happen between the pore water and carbonate minerals such as calcite or dolomite. More inert silicate minerals in the rock will react much more slowly – perhaps over hundreds to thousands of years.

You've noticed the uncertainty over the periods of times over which reactions occur? This is the bad thing about solubility and mineral trapping. We don't know enough about how long it will take to trap CO_2 in these more complex ways. This leads to a lack of confidence which influences investors who might finance big CCS projects. If we don't know in detail what happens to CO_2 over the long term then investors get nervous. More of this later.

To try to understand the rates of reaction, geological chemists are doing two things: building computer models for simulations and experimenting in laboratories. I'll explain some of their results.

These four odd looking squares (Figure 4.22) are actually photographs of an experiment that are again part of a 'film' that shows the dissolution of CO_2 into saline water. The squares are 'stills' which show the dissolution at different times.

The equipment was built specially and is basically a very thin square tank – like a single double-glazed window. Into the tank has been poured saline water which has been treated to go brown when it dissolves any CO_2. The tank is topped up with CO_2 gas.

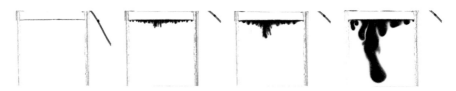

Figure 4.22 Photographs of an experiment that shows CO_2 dissolution. The first square shows the experiment after 1 min 45 s; the second after 4 min 30 s; the third 7 min 30 s; and the last 30 min. Courtesy Chris Rochelle. BGS copyright NERC.

The photographs were taken at times from the beginning of the experiment. The first square shows the experiment after 1 min 45 s; the second after 4 min 30 s; the third 7 min 30 s; and the last 30 min. A few things are interesting and important. One is that the CO_2 dissolves in the saline water quite quickly and begins to sink. It sinks because the CO_2-laden water is denser than the ordinary water. This is important because it indicates that CO_2 is going down rather than up. The second interesting thing is the shape of the descending CO_2-laden water which is in the form of bulbous 'fingers' that point downward.

The thing that's not so good about the experiment is that it's not very like the real situation below ground. For a start there is no rock in the tank. In the underground, these reactions would take place within connected pore spaces, not within free fluids. Also the experiment was done at room temperature and pressure, not anything like what would be expected at one kilometre below the surface. The experimenters also used gaseous CO_2 rather than supercritical CO_2. So the experiment has a lot of drawbacks − but it's a start.

Interestingly the experiment does bear out the patterns of dissolution that are predicted by simulations using computer models. One is shown below (Figure 4.23).

These simulations also need some explanation. There are two pictures showing different times. These times are actually dates: 2070 and 2270 − and the pictures show the predicted state of the plume of CO_2 at Sleipner on these dates pretty far off into the future. There are two coloured images in each picture. The left side shows 'undissolved' CO_2, and the right shows CO_2 in solution. The colours represent the amounts of CO_2: red is high concentration, blue is low.

In the 2070 picture (showing what the Sleipner plume might look like 60 years from now) there is quite a lot of undissolved supercritical CO_2. The main concentration of it is right under the seal (which is not shown). There are scraps of it also left under thin impermeable layers in the Utsira Sand. On the right in the same picture is the dissolved CO_2. There is a lot more of this because 60 years hence, much of the CO_2 at Sleipner will have dissolved. Much of the dissolved CO_2 is under the seal but you can see that some of it is descending from the seal in 'fingers' rather like in the experiment in Figure 4.22. By 2270, there's very little undissolved CO_2. Most of it is in solution and much

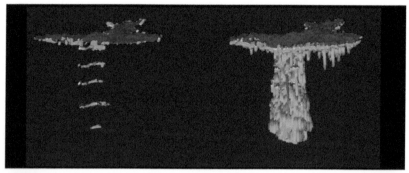

2070

2270

Figure 4.23 Simulations of the patterns of CO_2 dissolution at Sleipner. From Chadwick et al. (2008).

of it is descending in fingers from the seal because CO_2-laden solution is denser than the surrounding pore water. Some of it is sitting at the bottom of the Utsira Sand.

Some geochemists say dissolution and mineral formation within reservoirs loaded with CO_2 will take thousands of years. Some suggest that the seal and overlying rock might cause mineralization as well so trapping could go on above the reservoir even if it leaked.

STORAGE SPACE

Up to now we've looked how CO_2 behaves in rock. But we still need to answer a simple question. How much space is there for CO_2 — and is it enough to deal with our problem of climate change?

Geologists and organizations like geological surveys have spent hundreds of years mapping out rock layers and estimating the reserves of useful things in them like coal, metal ores, oil and gas. So it's only natural that they should turn to CO_2 storage space in rocks. There have been a lot of studies and a surprising amount is already known for example the 'Carbon Sequestration Atlas of the United States and Canada' published by the US National Energy Technology Laboratory (NETL) suggests that there is enough space in rocks in the US and Canada for a hundred years of CO_2 at present rates of emission. A similar figure for Western Europe is estimated by the European Union 'Geocapacity Project'.

We've looked at the basics of storage: porosity and permeability, but how is the capacity of a particular rock layer for CCS calculated? You'll know from the early part of this chapter that the storage capacity is not really to do with the volume of the rock layer. It's more about the space between the particles – so for example rock salt which I mentioned before as being crystalline with no pores at all would have no natural storage capacity, even if the layer was very thick and very widespread.

It just so happens that a recent large study on the storage capacity of offshore Britain has just been completed. The study, known as the United Kingdom Storage Appraisal Project (UKSAP), published the method it used to calculate the capacity and I'll explain it here.

The project geologists first identified deep saline aquifers and classified them into whether they were 'closed' or 'open' in relation to pressure. They also looked for traps within the deep saline aquifers. Then estimates of the porosity were used to work out the pore volume that might be accessible to CO_2 if it were injected. The geologists also modelled the rocks and did computer simulations of CO_2 flow rather like the ones I've shown you in the previous pages. The main reason for this was to see what effects flow might have on the capacity that was estimated. Not forgetting the ready-made depleted oil and gas fields, the geologists collected production and injection data from the oil and gas companies that operated the fields and calculated the amount of CO_2 that could be stored based on simply substituting the gas or oil with CO_2. The quality and security of the caprocks or seals of the various deep saline aquifers and depleted oil and gas fields were then checked. Finally the rough costs of using a particular storage trap or deep saline aquifer when it's linked with a power station or within a cluster were calculated.

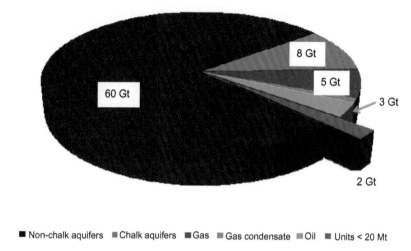

■ Non-chalk aquifers ■ Chalk aquifers ■ Gas ■ Gas condensate ■ Oil ■ Units < 20 Mt

Figure 4.24 British offshore storage space for CO₂. From presentation by David Clarke ETI (2012).

What were the results? The simple answer to the question of how much CO_2 could be stored in the rocks in offshore Britain is 78 giga-tonnes or 78 thousand million tonnes. This number is very large and has a lot of caveats attached to it, but it means – to go back to my local power station – that there is enough room for 7800 years of CO_2 from Ratcliffe-on-Soar. But of course this is a huge simplification because it would only be practical to store Ratcliffe-on-Soar's CO_2 nearby in the offshore – not for example in the northern North Sea.

The diagram (Figure 4.24) shows how this figure of 78 gigatonnes (Gt) is broken down. The largest proportion by far (60 Gt), 'non-chalk aquifers', are mainly deep saline aquifers of sandstone, like the Utsira Sand. The chalk of southern England that everyone is familiar with from the white cliffs of Dover also provides storage space deep under the seabed of the southern North Sea. Depleted fields of oil, gas and condensate (a substance half way between oil and gas) provide about 10 Gt together.

Where is the storage? The UKSAP project identified the best storage in the northern, central and southern North Sea (Figure 4.25), but also a few places in the east Irish Sea and the English Channel. You can see by the size of the blue and red open circles (blue = closed or confined deep saline aquifers; red open deep saline aquifers) that deep saline aquifers provide the largest storage potential.

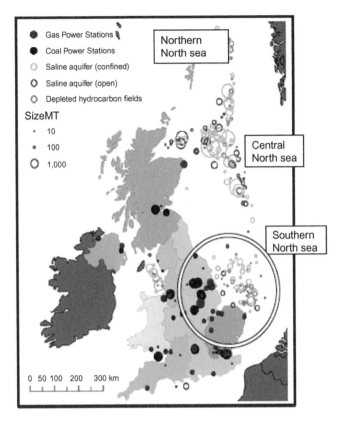

Figure 4.25 Location of British offshore storage space for CO₂. From presentation by David Clarke ETI (2012).

By joining up the dots of the power stations and other point sources and the open circles you can start to build your own CCS clusters. The most obvious is still the Humber, using storage in the southern North Sea. Another might be centred around the Firth of Forth in Scotland using the northern North Sea, or one centred on Liverpool and the Dee estuary using the east Irish Sea.

CO₂ STORAGE IN INDIA, CHINA AND SOUTH AFRICA

But Britain doesn't really matter in the global CO₂ game. With emissions of about 450 million tonnes per year Britain is dwarfed by the emissions of say the United States. Also as we saw in the first chapter, India, China and South Africa have emissions that are very likely to rise because of increasing coal use. In the IEA New Policies Scenario, for example, India will become the world's second largest consumer of coal (after China) by around 2025. By 2035 it might be using 880 million tonnes of coal per year.

It's right that rich countries like Britain and the United States do something about their emissions, after all they have the wealth and technical ability — and of course they are responsible for much of the historical emissions that have increased our CO_2 levels from 260–280 ppmv immediately before the industrial revolution to 395 ppmv today. There may even be some new business in developing the technology of CCS first. But it's clear that for CCS to make a difference, CO_2 storage has to be done in India, China and South Africa, as well as all the other emerging nations. Let's look at how feasible this is.

India

The CO_2 storage potential of India, like Britain's, is in deep saline aquifers and depleted oil and gas fields. India is an enormous country and so far assessment of capacity has been very crude. A recent study suggested that the storage capacity of depleted oil and gas fields is estimated to be between 3.8 and 4.6 Gt. An example of a depleted gas field that might become available is the Bombay High field which is offshore from the city of Mumbai and might offer a focus for a CCS cluster, however a simple estimate of the storage capacity of the Bombay High field gives a capacity of only about 600 million tonnes of CO_2 which is not enough to take large emissions for a long period of time. The same story is reflected all across India.

There appears to be lots of potential in the deep saline aquifers of the offshore basins in India though their huge size and the little that's known about them has prevented geologists from providing a figure for storage. The best areas seem to be offshore Gujarat and Rajasthan, and perhaps Assam (Figure 4.26).

China

Like India, China has a mixture of depleted oil and gas fields and deep saline aquifers that could be used for CO_2 storage. These are scattered amongst sedimentary basins (Figure 4.27). You might remember from earlier in the book that sedimentary basins are saucer-shaped depressions made of layers of sedimentary rocks going down quite deep.

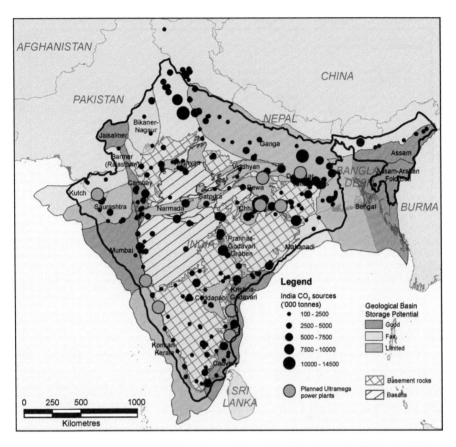

Figure 4.26 The CO₂ storage potential of deep saline aquifers in India. The rocks are classified as Good, Fair or Limited. The 'Good' deep saline aquifers have been shown to have once contained oil or gas and offer good reservoir and seal quality at depths >800 m over a large part of the basin. 'Fair' deep saline aquifers have some seals at depths >800 m and some trapping structures. 'Limited' areas have not much reservoir or are not sealed. Presentation given by Dr. Fatih Birol© OECD/IEA 2011.

In the Songliao Basin for example there are two oil field areas known as Daqing and Jilin, and a deep saline aquifer – the Qingshankou Formation. The Daqing area is thought to have about 593 Mt of CO_2 storage space in seven hydrocarbon fields after they are fully depleted, whereas the Jilin area has 102 Mt space in 5 hydrocarbon fields. The Qingshankou Formation has a bigger capacity of estimated at 692 Mt.

But the Songliao is just one of many basins. In general China storage capacities in oilfields are small compared to emissions from power stations – this is because the reservoirs are quite complex with

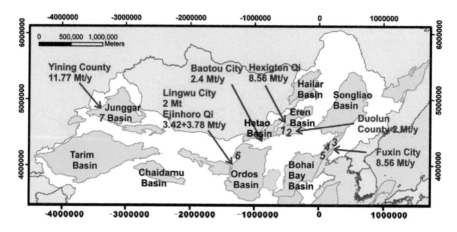

Figure 4.27 Some of the point sources and sedimentary basins of China. The point sources in red are listed accord-
ing to size in millions of tonnes per year (Mtly). The sedimentary basins are in green. Red numbers indicate smal-
ler sub-basins: 1 Manite sub-basin, 2 Tengger sub-basin, 3 Xialiaohe Depression, 4 Shuguang Oilfield, 5
Huanxiling Oilfield, 6 Wulanmulun Zhen, 7 Yining County. From Zeng et al. (2013).

porosity and permeability that isn't very high. To use this pore space
well, engineers will probably have to drill a lot of injection wells which
would be expensive. This probably means that deep saline aquifers will
be relatively more important in China than elsewhere, and so their
capacity needs to be urgently assessed. Not until this is done will a reli-
able figure for national Chinese CO_2 capacity be available.

South Africa

Recent reports suggest approximately 150 Gt of storage space in three
storage types (deep saline aquifers, unmineable coal seams and
depleted oil and gas reservoirs) of which about 98% is offshore. Most
of the space is in deep saline aquifers, with only limited storage potential in depleted oil and gas fields. Unfortunately these basins are offshore and far removed from the main CO_2 sources in South Africa.

FULL CIRCLE: STORING CO_2 IN COAL

In my discussion of the CO_2 storage capacity of South Africa I didn't
mention that South African geologists have been looking at the storage
potential of their coalfields. Their first estimate suggests that these
could store 1 Gt of CO_2. This is fascinating for a CCS geologist interested in the carbon cycle because the process would literally be putting
the carbon back where it came from!

How does storage in coal work? It's quite different from storage in sandstone where CO_2 is injected into the spaces between the particles of the rock. But coal — as I explained earlier in the book — isn't like other rocks. It's made of the compressed and chemically-altered remains of ancient plants. It doesn't have pore spaces like sandstone. Most of the space in coal is in cracks called cleats. Heat and pressure cooks up the old plant material and forms methane gas which fills the cleats, and the gas adsorbs onto the surface of the coal rather than floating free in the cleats. It can be released when the coal is disturbed or mined and the released gas is known to miners as *firedamp* or coal bed methane. The cleats are sometimes also filled with water.

So injection of CO_2 into coal — like injection into sandstone — can only happen if the CO_2 pushes the water or methane out of the way. In fact CO_2 adsorbs more readily onto the surfaces of coal in cleats and so displaces the methane. This makes injection feasible but also has a side effect that might be valuable — because if the methane is displaced by CO_2 it could be collected and used as fuel. This might be a way to gain some value from coal that is too deep to ever be mined, because in the process of storing CO_2 you produce something you can use. However, of course methane is a fossil fuel and would be burnt to produce more CO_2.

IS THERE ENOUGH STORAGE SPACE, THEN?

It's too early to say at the moment. The most sophisticated estimate of storage space — from Britain's UKSAP project — suggests that there are 78 gigatonnes or 78 thousand million tonnes of storage space in the offshore of Britain, mainly in the North Sea. This is an enormous capacity, and some geologists, engineers and planners see the North Sea as a great British asset, an area that might not only store CO_2 from British point sources but also from countries further afield, like Norway, Holland, France and Germany. Britain's North Sea oil and gas production is declining and CCS might be a way to maintain the momentum of industry.

But elsewhere the picture isn't so clear. The research hasn't been done yet. Our knowledge of storage in emerging countries like China, India and South Africa is sketchy. To be sure that CCS is the answer in countries that really matter we have to work much harder at assessing their capacity.

BIBLIOGRAPHY

Chadwick, A., Arts, R., Bernstone, C., May, F., Thibeau, S., Zweigel, P., 2008. Best practice for the storage of CO_2 in saline aquifers — observations and guidelines from the SACS and CO_2STORE projects. British Geological Survey, Nottingham, UK, 267pp. (British Geological Survey Occasional Publication, 14).

Council for Geoscience South Africa, 2010. Technical report on the geological storage of carbon dioxide in South Africa.

Gammer, D., Green, A., Holloway, S., Smith, G., The UKSAP Consortium 2011. The Energy Technologies Institute's UK CO_2 Storage Appraisal Project (UKSAP). SPE Paper Number 148426.

Holloway, S., 2007. Carbon dioxide capture and geological storage. Philos. Trans. R. Soc. A 365, 1095–1107.

Holloway, S., Pearce, J.M., Ohsumi, T., Hards, V.L., 2005. A review of natural CO_2 occurrences and releases and their relevance to CO_2 storage. BGS External Report CR/05/104, 117 pp.

Smith, D.J., Noy, D.J., Holloway, S., Chadwick, R.A., 2011. The impact of boundary conditions on CO_2 storage capacity estimation in aquifers. Energy Procedia 4, 4828–4834.

Stephenson, M.H., Rochelle, C., 2010. Have your coal and burn it. Educ. Chem. 47, 23–25.

Studlick, J.R.J., Shew, R.D., Basye, G.L., Ray, J.R., 1990. A giant carbon dioxide accumulation in the Norphlet Formation, Pisgah anticline, Mississippi. In: Barwis, J., McPherson, J.G., Studlick, J.R.J. (Eds.), Sandstone Petroleum Reservoirs. Springer, New York, NY, pp. 181–203.

Zeng, R., Vincent, C., Tian, X., Stephenson, M.H., Xu, W., Wang, S., 2013. New potential carbon emission reduction enterprises in China: deep geological storage of CO_2 emitted through industrial usage of coal in China. Greenhouse Gases Sci. Technol. Available from: http://dx.doi.org/10.1002/ghg.

CHAPTER 5

Will It Leak?

By 2025, CO_2 from Ratcliffe-on-Soar power station will perhaps begin to be injected into the Bunter sandstone offshore under the North Sea. How will we know that the CO_2 is going into the right reservoir, and how do we know that it will stay there? What is the risk of CO_2 coming back to the surface, and what would it do to the North Sea if CO_2 was bubbling from the seabed? We saw at Sleipner that it's possible to actually see CO_2 collecting in sandstone almost 1 km below the bed of the North Sea. We can see how the shape of the plume has changed over time, and we can calculate how much CO_2 is there. So imaging CO_2 is not so difficult. Using the same seismic cross sections, we can also scour the rock layers for faults so that we can avoid them. We can avoid any old oil wells that might leak. If we manage the 'CO_2 store' properly it won't leak, but if it does we can detect leaks at sea and on land, and we can fix them. But in the long years after the store is full and officially closed someone has to take charge. We know that CO_2 is less and less likely to leak the longer after closure you go — because of dissolution and mineralisation — but even then it's not reasonable for a CCS operating company to be responsible for the store for hundreds or thousands of years after it has closed. Companies don't last that long anyway! The fact is that governments and state institutions — which do last longer — will probably have to take ownership of stores in the long term.

Keywords: *CO_2; leak; monitoring; regulation*

Returning Carbon to Nature: Coal, Carbon Capture, and Storage. DOI: http://dx.doi.org/10.1016/B978-0-12-407671-6.00005-7

To answer whether CO_2 can leak we can look at an example. You might remember that in Chapter 4 I introduced you to the Bunter sandstone, a rock layer that underlies a lot of mainland England and also the North Sea and Irish Sea. CCS scientists have their eye on it as a good place to put CO_2 because it's very porous and permeable, and because it often forms convenient arches deep below the surface that could be used to trap buoyant CO_2. In Figure 4.11, I showed you a small part of the Bunter sandstone as it appears deep underneath the seabed offshore from the Humber estuary in eastern England. This is another reason why CCS geologists are interested in the Bunter sandstone – because it's close to the Humber cluster of big power stations. This layer of rock may be the first in Britain to be connected to full-chain CCS, and could in the future receive CO_2 from Ratcliffe-on-Soar power station.

Figure 4.11 is a picture from a 3D computer model, and the image has been squashed sideways ('vertically exaggerated') to emphasize the ups and downs of the layer. Figure 5.1 shows more of the rock layers under the North Sea, including layers above and below the Bunter sandstone. Similar to Figure 4.17 this is a seismic cross section and again it's been highly vertically exaggerated so that the arches and troughs appear much more pronounced.

The diagram shows some layers very clearly and others less clearly. Our Bunter sandstone layer which is between the green line and the red line is down quite deep and is part of a line of arches and troughs.

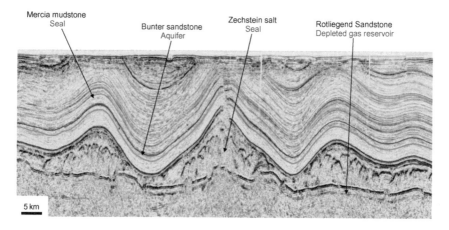

Figure 5.1 A seismic section in the southern North Sea showing the Bunter sandstone and the layers above and below it. The Bunter sandstone is interpreted to lie between the green and red lines. Courtesy of WesternGeco.

Above it is a mudstone layer (a very fine grained sedimentary rock) known as the Mercia mudstone that acts as a seal or caprock for gas fields in parts of the North and Irish seas. CCS geologists think that the Mercia mudstone will be an excellent seal for CO_2. Above the Mercia mudstone are other seal layers and reservoirs right up to the seabed which is the faint brown line on the cross section.

Below the Bunter sandstone are other layers. The rather disorganized layer is the Zechstein salt which was deposited at the bottom of a salty sea, not unlike the North Sea in position and shape, at a time when Europe was very arid, about 250 million years ago. You might remember from Chapter 4 that salt is very impermeable because it's crystalline and therefore has no gaps or space to allow liquids or gases to flow. So the Zechstein salt is a seal. The layer under that – just below the yellow line – is the Rotliegend sandstone which is a natural gas-bearing layer. In parts of the North Sea, for example in the area shown in Figure 5.1, the Rotliegend sandstone has depleted gas fields because gas has been extracted.

So the figure shows both the types of storage that I introduced in Chapter 4 – deep saline aquifers or formations, and 'ready-made' storage. Knowing what we do about the geology, how might CO_2 stored either in the Bunter sandstone or the Rotliegend sandstone leak upwards into the rocks above, or into the sea and eventually into the atmosphere? If we can identify the way it could happen, can we say how likely it is to happen, and what the consequences might be?

Embedded in the idea of how likely something is to happen – and how bad the consequences are – is the concept of risk. We all take risks – crossing the road, driving a car or a motorcycle, smoking a cigarette – and when we take risks, probably subconsciously we weigh up the risk. In weighing it up we're really balancing risk and reward. We're asking ourselves whether it's worth it.

We'll use the same method here in working out whether CCS is worth the risk.

HOW LIKELY IS LEAKAGE

The complicated diagram (Figure 5.2) shows the main ways that geologists think that an artificial store of CO_2 could leak and the ways that

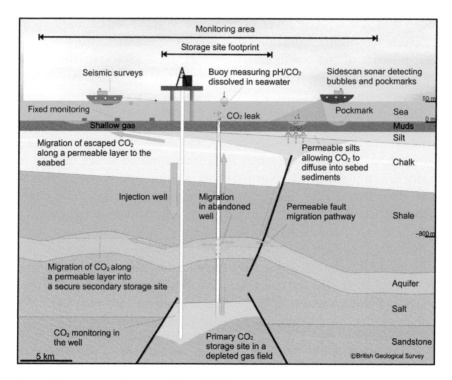

Figure 5.2 Ways that an artificial store of CO_2 could leak and ways that you might detect the leak. Courtesy of the BGS. Courtesy Michelle Bentham. BGS copyright NERC.

you might detect the leak. The main way is thought to be through abandoned or old oil or gas wells. This might seem odd to anyone that hasn't worked in the oil and gas industry, but when wells stop producing anything useful, they are usually filled in and blocked. In this process, the tube is removed from the well, and sections of the hole are filled with cement to seal the gas- or oil-bearing part from the water-bearing part, and to plug the well at the surface. The surface around the top of the well is excavated and a steel cap is welded in the top of the hole.

Figure 5.2 shows what could happen if an abandoned well leaked CO_2 because it wasn't sealed or capped properly.

The well is shown in the centre of the picture. Like the injection well (the well used to inject the CO_2), the abandoned well penetrates the CO_2 storage site which is sealed by a layer of salt on top. Perhaps the CO_2 is able to get into the bottom of the abandoned well because there are cracks in cement or because the cement has come away from the wall of the well. If the well is completely open right to the top, then it's possible that CO_2 could leak straight into the sea. It's also possible that it could collect in a porous and permeable layer on its way up. In Figure 5.2, for example, it finds its way into a deep saline aquifer. In the diagram, an impermeable shale layer seals the CO_2 in the aquifer, but it could migrate sideways and upwards – remember it's buoyant – into an arch. In this case there would be no risk of further migration because the shale acts as a secondary seal. One of the rather good things about rock layers, in the North Sea for example, is that they tend to occur in alternations of sandstone and shale – which is why the area is so good for oil and gas. This also means that there are 'stacked' seals which would act as stopgaps in the event of a leak deep down.

I treated faults as impermeable barriers in Chapter 4, for the purposes of explaining a 'closed' CO_2 reservoir. Usually they are exactly that. This is because the two sides of the fault are jammed so close together, and because they are often filled with impermeable minerals. But we know that sometimes faults can also allow gases and liquids to flow along them. It's also possible that when you pump high pressure gas or liquid into a rock that has an old fault, you can open the fault up and make it move. I'll explain more about this later, but for now imagine that a fault can allow CO_2 to move upwards. In the diagram above, the fault allows CO_2 to get into a porous and permeable chalk layer. From there it has got into the seabed mud and silt and is slowly bubbling into the seawater.

What are the chances of these leaks happening? The main way that we can reduce the risk is by understanding the rock layers that we want to use to store CO_2. If we know that there is an abandoned well then we'll have to find out all about it, including its condition, age and the way in which it was plugged. Databases exist that give information about abandoned wells in the North Sea. Obviously if there is any doubt about a particular well, then geologists would look for somewhere else to store CO_2. The same goes for faults. In the case of the diagram the faults that confine the lower depleted gas field must have been seals or the natural gas that was in the structure would have escaped. The geologists and engineers would have to be careful that pressurizing the reservoir with CO_2 would not open the faults even by a minute amount. This may sound difficult but in fact oil companies are expert in managing the pressure of gas reservoirs like this mainly because the gas is the company's bread and butter, and the last thing they want to do is let it escape.

If we don't know whether a fault is a seal or a conduit, the best thing is to avoid it. You might ask — how do you do that? Well faults can be seen on seismic cross sections. In fact large ones are pretty obvious. If you look at the middle arch in Figure 5.1 you can see a fault that has displaced rock layers. The Bunter sandstone has been displaced downwards to the right. This arch would not be a wise place to store CO_2.

But this kind of detailed work characterizing a possible storage site isn't easy or cheap. Seismic surveys are very expensive; drilling test boreholes to understand rocks before you inject CO_2 is also expensive. But this money will have to be spent so that risk can be minimized. If you avoid faults and poorly plugged abandoned wells the likelihood of leakage is very low. The Intergovernmental Panel on Climate Change (IPCC) puts this risk in a particular way by saying that in *appropriately selected and managed reservoirs* the fraction of CO_2 retained is *very likely to exceed 99% over 100 years*, and *likely to exceed 99% over 1000 years*. Essentially this means that less than one per cent of the CO_2 stored would be lost over a century or a millennium.

IMPACTS

As I said earlier, risk can be thought of as having two components: the likelihood of something going wrong and the impact of it going wrong. Some people would go as far as to say that risk is likelihood *multiplied* by impact. So we need to consider the impact of CO_2 injection and leakage.

The main impact of CO_2 leakage underground is contamination of water wells with CO_2. In Britain where CO_2 is only likely to be stored under the seabed this will not be a problem, because there is no fresh water supply from rocks under the sea. In places where CO_2 storage is contemplated onshore, however, what would be the impact of CO_2 leakage? In fact the water would become carbonated — not unlike commercial carbonated bottled water. This in itself would not be dangerous, but it would be inconvenient — and water with dissolved CO_2 — which is more acidic than ordinary water — would be more corrosive to pipes and tanks.

What if CO_2 bubbles into seawater from a rock layer below? Because the conduit is likely to be a 'point source' the impact will probably be over rather small areas. The most obvious impact would be acidification of the seawater because CO_2 dissolves in seawater. This we would see as a lowering of the pH of the seawater. Although it's uncertain exactly how this would affect sea life, because we neither know the effects of pH decrease nor the resilience of sea life to pH change, there would be serious change to the ecosystem which would likely affect human activities such as fishing. As an example, the coccolithophore – a very abundant phytoplankton that armours itself with a calcium carbonate skeleton – probably would not be able to make its carbonate skeleton so well in low-pH seawater. Unknown effects on primary producers like coccolithophores – the source of much of the ocean's food – would feed up into higher sea life. It's also unknown what the effects of a decrease of pH would be on big ocean processes such as carbon and nitrogen cycling.

We know that there are tides and currents in the sea that would be capable of moving low-pH water around. What would the pattern of acidification look like? It's possible to model this with computers; Figure 5.3 depicts a series of maps which show simulated leaks in two possible seabed locations in the North Sea. In the simulations the modellers assumed large continuous leaks. For the simulation, the rate of leakage of CO_2 was set at five times the rate of CO_2 injection at the Sleipner CCS site – so these simulated leaks are very large. The amount of acidification is shown by the amount of decrease of pH in tenths of a pH point.

Let's look at the first leak in the north of the North Sea somewhere offshore of Aberdeen (the sequence of maps labelled (A). The first map shows nothing but after a few months (by an arbitrary date of 15 March), the area of reduction of pH is shown. The reduction is one tenth of a pH point. Because of tides and currents and continued leakage, the area of reduced pH grows and changes in shape. In the sequence of maps labelled (B) the effects of a leak of the same size (same rate of CO_2 escape) seem to be larger. The area of pH reduction is wider, and there are stronger reductions in pH also.

What about impacts on land? The effects of high concentration of CO_2 on humans are quite serious and are well known from studies of submariners. From normal atmospheric levels (about 0.03%) to 3% CO_2, people breathe faster and get headaches. Levels of 15% are fatal

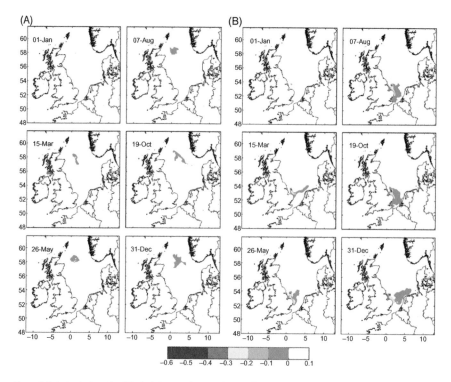

Figure 5.3 A simulation of a CO_2 leak in two places in the North Sea (A and B) From Blackford et al. (2009).

to humans. Higher levels of CO_2 are tolerated by many soil organisms and burrowing animals for obvious reasons. Plants have very wide tolerance of CO_2 levels, some being able to thrive in conditions where CO_2 is at 50%.

In some parts of the world quite high levels of CO_2 are encountered daily, for example in parts of Italy. Figure 5.4 shows a natural leak of volcanic CO_2 near Rome's Ciampino airport. The amount of CO_2 being produced is more than the vegetation can tolerate, and so the area is barren. The leak poses no danger to humans in the open air because CO_2 concentration above ground level rarely gets to dangerous levels, but the basements of buildings in the area have to be carefully monitored. Parts of Naples have the same problem.

So the impact of CO_2 leaks, though somewhat uncertain, is probably quite serious on land and at sea. Even if leaks are very unlikely because we can avoid the kind of structures that cause them (like faults and leaking wells), we want to avoid them happening at all − or if

Figure 5.4 Ciampino CO_2 leak, located in a populated area in Rome. Photo taken from a plane landing at Rome Ciampino airport. The red ellipse outlines the place where the natural CO_2 is leaking. Courtesy SCCS/Roberto Bencini.

they do happen we want to be able to detect them and fix them before they do any damage. So can we detect leaks?

DETECTING LEAKS

Figure 5.2 shows some of the methods that could be used to detect leaks in the sea above a CO_2 store including seismic surveys, fixed sea-bed sensors, pH sensors on buoys and sidescan sonar that can detect bubbles. There are also physical signs that leaks produce, known as pockmarks, which form in soft sediment on the seabed where gas is leaking. These can be used as evidence of leakage if they appear after a CO_2 store has been established.

On land, detecting leaks is more difficult because there are no bubbles. However scientists are working on the problem by concentrating on places where we know that CO_2 naturally leaks to the surface, for example in volcanic areas of Germany and Italy.

The crater lake Laacher See in Germany is an example. It was formed in a volcanic eruption about 13,000 years ago. Despite its tumultuous origin the lake is now a quiet beauty spot on whose banks is the Benedictine monastery of Maria Laach Abbey. Although the 13,000-year-old eruption is long over, the volcano − deep below the ground − is still fuming, and natural CO_2 is coming to the surface in the lake and on its grassy banks.

In the last few years the quiet solitude of the lake has been disturbed by noise from some outlandish machines, one of which is pictured in Figure 5.5. The quadbike in the picture is being used by CCS scientists to detect and measure CO_2 coming to the surface through cracks and fissures.

The detector on the front of the quadbike is an 'open path laser gas analyser' which can measure to an accuracy of 5−10 parts per million of CO_2 on top of the CO_2 that's already in the air, so it can detect very small leaks. The machine takes a reading every second, and the position of the reading is located using GPS mounted on the quadbike. The output of the machine is very interesting. Although there are actually leaks of CO_2 occurring in the meadow that you can see in the

Figure 5.5 CO_2 detection equipment mounted on a quadbike being used on the banks of Laacher See in Germany. Courtesy Dave Jones. BGS copyright NERC.

photograph, they don't have much of an impact on the plant and animal life, unlike at Ciampino airport. In fact if surveys like the one carried out by the quadbike hadn't been done, we wouldn't know just how much CO_2 is leaking at Laacher See and the variety of leakage points.

In Figure 5.6, the green line is a map of the zigzag track of the quadbike, and the colours show CO_2. The red and yellow points show places where CO_2 is coming to the surface from the ancient volcanic rocks below. The leakage points are very small — only a few meters wide. Figure 5.7 shows many tracks of the quadbike superimposed on an aerial photograph of the banks of the lake. You can see the tracks as a faint zigzag black line.

The colours superimposed on the photograph show atmospheric concentration of CO_2 in the air just above the ground. The higher

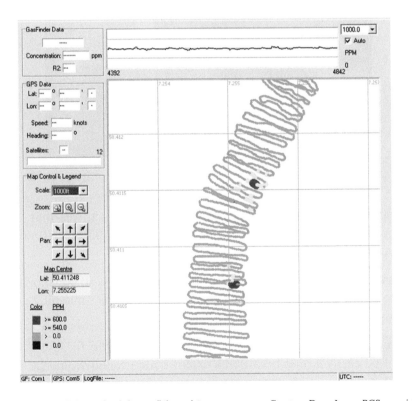

Figure 5.6 Map of the track of the quadbike and its measurements. Courtesy Dave Jones. BGS copyright NERC.

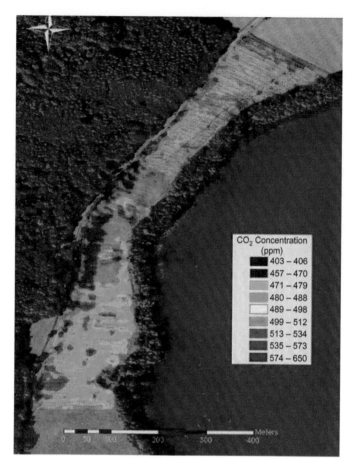

Figure 5.7 Natural CO₂ leakage on the banks of Laacher See. Courtesy Dave Jones. BGS copyright NERC.

concentrations are shown by warmer colours. You can see that there are several areas mostly less that 100 m wide that indicate that natural volcanic CO_2 is leaking up to the surface.

This shows that we can detect and measure leaks, so we should be able to tell if something is going wrong. But if there is something going wrong, can we do something about it?

An old well leaking CO_2 is quite straightforward to fix, but a fault cannot be easily sealed because its extent underground would be diffi-cult to measure. If leakage occurred, as in Figure 5.2, and CO_2 was found to have leaked into a secondary storage site and to be on its way up to shallower rock layers it would be necessary to stop injection to reduce the pressure. This might be enough to prevent further leakage. If there was clear sign of a water aquifer being contaminated with CO_2, then water can be extracted and cleaned and then returned. Similarly any leaked CO_2 can be pumped out. This will cost the operating company a lot of money and within the European Union Emissions Trading Scheme (ETS) they would also have to pay a penalty for the release of the CO_2 into the atmosphere. More about this described later.

THE BIG LEAK

It's worth stepping back at this point though, because we have to consider leaks in CO_2 in the context of carrying on without a good plan for CO_2 emissions reduction. The truth is that CO_2 is 'leaking' directly into the atmosphere from power stations all around the world right now because we're not making an effort to capture the CO_2. Rather than worrying about small leaks that might happen from a CO_2 storage site, we should try to see what effects our present 'big leak' might have on the environment, apart from the obvious problem of global warming.

More CO_2 in the atmosphere means that all the seas on Earth will become more acidic. The group that looked at the effects of point source leaks from CO_2 stores in the North Sea (Figure 5.3) also tried to model how the southern North Sea would acidify purely from contact with the atmosphere as it becomes more concentrated in CO_2. In their simulation which takes place over the next century (Figure 5.8) we see acidification increasing by over half a pH point in surface water over the whole of the southern North Sea. The effects of this change would clearly be very serious.

REGULATION

Several European countries are taking CCS very seriously, including Britain. The regulations that will govern CCS are being formulated right now. At the heart of the regulations for CCS in Europe is a common approach that was defined in a document known as the 'Directive on the Geological Storage of CO_2', which was published by the European Commission in 2009. This provides regulations for permanent CO_2 storage where the amount of stored CO_2 in total after the operation is more than 100,000 tonnes. The *Directive*, as it's

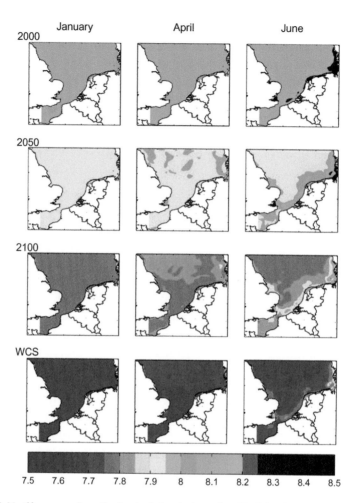

Figure 5.8 Monthly mean surface pH value simulations in the southern North Sea from predicted levels of CO_2 increase in the atmosphere. From Blackford and Gilbert (2007).

known, looks very closely at whether injected CO_2 is behaving as expected, whether any leakage is occurring, and if this leakage is damaging the environment or human health. Only if a CCS operator satisfies requirements will a company be allowed to keep a licence and a permit to store CO_2. The Directive also contains a condition that has to be satisfied to allow the company to close the site when it's full of CO_2: the company has to demonstrate that the storage site is evolving towards a *situation of long-term stability*.

What does this mean practically? Well let's go back to the Bunter sandstone in the North Sea. Let's imagine that a site has been chosen, and that some detailed geological work has been done to understand the site. The geologists have used seismic cross sections and information from nearby oil wells to understand where the reservoir and the seals are. They've also looked for faults in the seismic cross sections and checked for abandoned wells. If there are abandoned wells, the geologists have checked that they've been plugged properly.

The geologists will have to prove that they've done all this work to a high enough standard to convince the regulator that they can start to inject CO_2, in other words to get a permit. But the permit can be withdrawn if things look like they're going wrong. The company has to show that injected CO_2 is behaving as expected, as well as monitor migration or leakage and any damage to the environment or human health. How will the operator know that things are going well? How will the regulator check? First, the operator will probably use seismic cross sections in the same way that they've been used in Sleipner to image the CO_2 and track its flow direction (see Figure 4.19). If the CO_2 moves somewhere that the simulations didn't predict then this might be grounds for withdrawing a permit.

The pictures of the Sleipner plume from Chapter 4 were gathered at great expense by running seismic surveys over the same place several times with a few years in between. This might be too expensive a way to monitor the movement of underground CO_2 in the future. It's also rather inconvenient to carry out a seismic survey on land because loud sounds or vibrations are generated at the surface from big trucks. This means that other ways of monitoring CO_2 movement are being developed including detecting tiny changes in the gravitational field in the area of the CO_2 storage or using seismic detectors installed in nearby monitoring boreholes. It might also be realistic to have geochemical sensors in monitoring boreholes that detect pH changes in pore water as CO_2 is injected.

What about that last stipulation in the Directive: that the company has to demonstrate that the storage site is evolving towards a *situation of long-term stability*? This is quite a thorny issue and perhaps the largest scientific problem in CCS. What happens to CO_2 in the long term, and who is responsible for the store if it goes wrong a long time into the future?

What we have to remember is that CO_2 is probably going to be down in the rocks for a very long time, unless we find some use for large amounts of it. You'll know from Chapter 4 that stored CO_2 will first dissolve in pore water and then descend through the reservoir because it will be denser than ordinary pore water. At some time after that it's likely that the acidic pore water will react with the rock particles that form the reservoir to make various carbonate minerals. The descending CO_2-laden pore water is good because the CO_2 that we've spent so much time and money injecting is going downwards rather than up towards the surface. But the 'holy grail' of CCS is the formation of carbonate minerals because CO_2 locked up in a solid mineral will never leak. As I said in Chapter 4 we don't know the exact timescale over which this will happen. Perhaps the geologists and the engineers will not be able to demonstrate that things are tending towards a *situation of long-term stability*.

Why is this a problem? This is mainly because the companies that operate the CO_2 storage site may not be around for hundreds of years, still less than thousands of years. Realistically, it's more sensible for the regulator or the government to take responsibility for the store after it's full and has reached long-term stability. This should not be onerous for the regulator because the store should cause no problems after it's stable. But it's not fair for the regulator (the government) to take on too much risk, particularly because if anything goes wrong then tax-payers' money will be used to fix it. On the other hand we can't expect the company to keep on operating the store long after they've closed it, waiting for it to reach a point of stability. It might go out of business because it's no longer making any money from its CO_2 disposal business (more about this later too!). If the risk is too big for the companies then they won't want to get involved in the CCS business. If the risk is too big for the regulator, then they will also want out.

We can illustrate this dilemma in Figure 5.9. This is a risk profile for CCS giving one scientist's view of how risk varies through the process from start of injection to end of injection – and long after injection.

In Figure 5.9 risk goes up from the time of injection until the end of injection, and then through pressure decrease and dissolution ('secondary trapping mechanisms'), risk starts to go down. Can we draw a vertical line across the graph where the handover from the operating company to the regulator takes place? It has to be at a point that doesn't frighten off the operating company and its investors, but at the

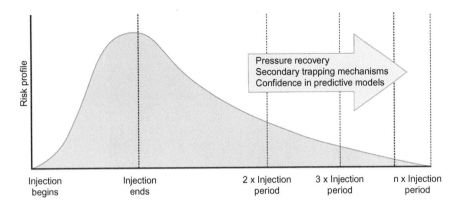

Figure 5.9 Where do we draw the line in the risk curve for CCS so that the balance of risk between the operating company and the regulator is correct? From Benson (2008).

same time has to be at a point where there isn't too much risk for the regulator — or for public opinion. This isn't an easy decision.

I've not mentioned public opinion about CCS very much yet. This is partly because the public don't know much about CCS except in places where it's being done, or where it's being planned.

The oil company Shell announced a project in 2007 to store the CO_2 by-product from an oil refinery under the town of Barendrecht in the Netherlands. After the announcement was made, local public opposition became very strong resulting in many public meetings and campaigns. Posters and images as in Figure 5.10 were used to marshal public opinion against CCS. Opposition became so strong that in November 2010 the Dutch Ministry of Economic Affairs, Agriculture and Innovation announced the cancellation of the project. Post mortems suggested that the company didn't provide enough information and didn't talk to local people enough about the project.

But there are also signs that public opinion can be favourable to CCS. Near the spectacular cliffs on the coast west of Melbourne in Australia is the Otway CCS test project where over 65,000 tonnes of CO_2 have been injected and stored in a depleted gas reservoir. The project was discussed with local people at the outset, and the researchers have been completely open about the objectives of the project and the aims of CCS as a whole — with the result that the work is generally accepted. There is even a sense of pride that the project has been so successful and attracted international interest.

Figure 5.10 A poster used to mobilize opposition to CCS in the Dutch town of Barendrecht. From http://www.
stichtingmilieunet.nl/andersbekekenblog/klimaat/shell-boos-om-fictieve-CO$_2$-ramp-in-barendrecht.html.

BIBLIOGRAPHY

Benson, S., 2008. Multi-phase flow and trapping of CO$_2$ in saline aquifers. (Paper No. OTC 19244). Proceedings of 2008 Offshore Technology Conference. Houston, TX, 5–8 May 2008.

Blackford, J.C., Jones, N., Proctor, R., Holt, J., 2009. Regional scale impacts of distinct CO$_2$ additions in the North Sea. Mar. Pollut. Bull. 56, 1461–1468.

Blackford, J.C., Gilbert, F.J., 2007. pH variability and CO$_2$ induced acidification in the North Sea. J. Mar. Syst. 64, 229–242.

Chadwick, A., Arts, R., Bernstone, C., May, F., Thibeau, S., Zweigel, P., 2008. Best Practice for the Storage of CO$_2$ in Saline Aquifers – Observations and Guidelines from the SACS and CO$_2$STORE Projects. British Geological Survey, Nottingham, UK, 267pp. (British Geological Survey Occasional Publication, 14).

Metz, B., Davidson, O., de Coninck, H., Loos, M., Meyer, L. (Eds.), 2005. Carbon Dioxide Capture and Storage IPCC. Cambridge University Press, UK.

CHAPTER 6

Accounting for Carbon

In the preceding chapters I've shown how carbon capture and storage (CCS) might reduce CO_2 emissions. The technologies for capture, transport and storage are already here, and are being demonstrated separately day-in day-out throughout the world. There are also a few projects that demonstrate part or all of the full chain going from capture to storage. In Sleipner, for example, a million tonnes of CO_2 per year is captured from a gas field and stored in the Utsira sandstone. This CO_2 is from natural gas that's been extracted from a layer below the Utsira sandstone which would otherwise have been vented to the atmosphere because natural gas with a lot of CO_2 can't be used on the gas supply grid. The Norwegian government set a tax of about US\$50 per tonne of CO_2 released and so the operating company Statoil avoided paying the tax by disposing of the CO_2 underground. In a sense the 'tax avoided' pays for the effort and technology involved in working out how to do CCS and in actually doing it. This may be the key to making CCS economically feasible. But what part should CCS play in emissions reduction? Can it provide all the reductions we need, or will we have to use it with other technologies? How much will we have to change our lifestyles? It's hard to predict the future but in this chapter I survey the landscape out to 2050.

Keywords: *CO_2; carbon cycle; carbon price; PETM; climate change*

There are other ways apart from carbon capture and storage (CCS) to reduce CO_2 emissions. How do we choose what to do? Do we go for lots of renewables (e.g. wind, tidal); do we go for lots of nuclear; do

Returning Carbon to Nature: Coal, Carbon Capture, and Storage. DOI: http://dx.doi.org/10.1016/B978-0-12-407671-6.00006-9

we choose mainly CCS and continue to use a lot of fossil fuels? These are obviously very complex questions to do with the level and cost of the technology, but also policy.

One way to guide our decision is to look ahead to sometime in the future and try to plan. Which of the technologies is most efficient at reducing overall emissions? How much CO_2 can be prevented from entering the atmosphere? How do we account for this CO_2?

A now famous article by Stephen Pacala and Robert Socolow in the journal *Science* in 2004 helped to concentrate the minds of policy makers and scientists. They simplified the problem into two questions – where are we going and where do we want to be? As to where we're going – well at the moment the world isn't doing much about CO_2 emissions: some countries are making efforts but others are not. What will this pattern of increase in emissions look like 10 years ahead, or 30 years ahead? This pattern is not unlike the 'business as usual' forecast and the 'Current Policies Scenario' of the IEA that I mentioned in the first chapter.

Where do we want to be? In their article and in later work, Pacala and Socolow settled on the idea of stabilizing carbon emissions at 7 billion tonnes per year. This, they say, will avoid the worst effects of climate change.

Before we move on I should point out that the word 'carbon' rather than 'carbon dioxide' is beginning to creep in. In fact in the policy debate about climate the two are used interchangeably – but they are quite different. In this book I've tried to keep to using emissions in terms of CO_2 for simplicity, but emissions of carbon alone are often spoken about. This is really just shorthand because solid carbon can't be emitted to the atmosphere and probably wouldn't harm it anyway. So we need a conversion figure to translate between emission as CO_2 and emission expressed only as 'carbon'. Carbon has an atomic weight of 12, whereas CO_2 has a molecular weight of 44. If emissions are expressed as tonnes of carbon then the amount has to be multiplied by $44/12 = 3.67$ to convert it to tonnes of CO_2. So Pacala and Socolow's stabilization of carbon emissions at 7 billion tonnes per year equals 25.7 billion tonnes of CO_2.

Pacala and Socolow say that carbon (and CO_2) emissions from fossil fuel burning are projected to double in the next 50 years (Figure 6.1).

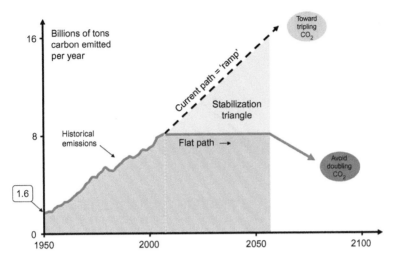

Figure 6.1 Carbon (and CO$_2$) emissions from fossil fuel burning are projected to double in the next 50 years. So there will be triple the atmosphere's CO$_2$ concentration compared to the level before the industrial revolution. From the Carbon Mitigation Initiative website, Princeton University http://cmi.princeton.edu/wedges/. Courtesy of the Carbon Mitigation Initiative, Princeton University.

This will mean that for 'business as usual' there will be triple the atmosphere's CO$_2$ concentration by 2055 compared to the level before the industrial revolution. Pacala and Socolow argue that if emissions can be kept flat over the next 50 years (orange line), it would be safer. The *flat path*, followed by emissions reductions later in the century, is predicted to limit CO$_2$ rise to less than a doubling and avoid the worst-predicted consequences of climate change. Keeping emissions flat for 50 years will require cutting carbon emissions by roughly 8 billion tonnes per year (or about 30 billion tonnes of CO$_2$) until 2055, stopping a total of 200 billion tonnes of carbon (or about 730 billion tonnes of CO$_2$) from entering the atmosphere. Pacala and Socolow call these 'carbon savings' the *stabilization triangle* (the yellow triangle in Figure 6.1; enlarged in Figure 6.2).

But the really clever thing that Pacala and Socolow did was to divide this stabilization triangle into a group of smaller triangles called wedges. Each wedge is a manageable amount of effort that could avoid 1 billion tonnes of *carbon* emissions cumulatively by 2055 (Figure 6.2). In other words it's an activity that on its own, if done between now and 2055, could stop a billion tonnes of extra carbon (or 3.67 billion tonnes of CO$_2$) from getting into the atmosphere. The activity could for example be house insulation so that less heating fuel is burned; or

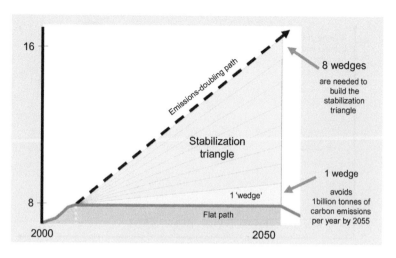

Figure 6.2 The stabilization triangle. Courtesy of the Carbon Mitigation Initiative, Princeton University.

it could be making cars more efficient; or it could it be wind power. In the opinion of Pacala and Socolow, the stabilization triangle could be achieved by making eight wedges. You can think of a wedge as a sort of scientific unit of CO_2 abatement.

Below are some of the wedges that they suggest (Figure 6.3).

Let's look at some of these in more detail. What do they mean by 'Fuel Switch', for example? Well if you compare gas power stations with coal power stations the former are cleaner in the sense that per unit of energy produced, gas power stations also produce rather less CO_2. If 1400 natural gas power stations were substituted for an equal number of coal-fired power stations then just this swap would save *one wedge* of CO_2 emissions. This is without having to resort to CO_2 capture and storage. The nuclear fission wedge is simple: triple the world's nuclear electricity capacity by 2055 and we'll achieve a wedge. In wind, the installation of 1 million 2 megawatt (MW) windmills to replace an equivalent amount of coal-based electricity generation would produce a wedge. In solar, the installation of 20,000 km^2 of solar panels by 2055 would produce a wedge. Perhaps simplest of all would be to eliminate tropical deforestation or plant new forests over an area the size of the continental United States. Both would produce a wedge.

CCS gets its own mention in Pacala and Socolow's work. They suggest that if CCS was done on coal power stations to a total of 800 GW

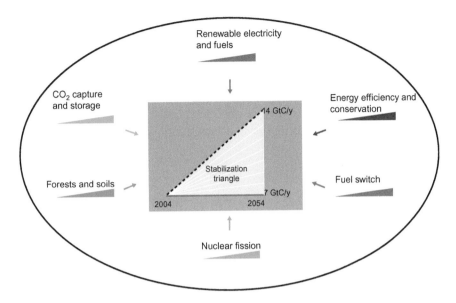

Figure 6.3 Types of wedges. Courtesy of the Carbon Mitigation Initiative, Princeton University.

capacity (about 200 large coal power stations) and the CO_2 was stored underground, then this would achieve a wedge.

Of course, I hear you saying it's much more complicated than that. Options like nuclear power, for example, have their own special difficulties. Some people don't like nuclear power because there are questions over safety and cost. Similarly the elimination of tropical deforestation is easy to say and very difficult to do.

But we can't accuse Pacala and Socolow of being naïve. What their work has done has shown that all the wedges they suggest could be achieved. Without going into detailed cost analysis and assessment of technology, all the wedges appear to be feasible. In other words human beings already have the right tools to achieve the stabilization. But we'll need a variety of wedges — there's certainly no one technology or change in behaviour that can do the job.

Let's now look at a more sophisticated model of emissions reduction. The '2050 Pathways' project was developed by the British Government to investigate different scenarios to reach an 80% reduction in greenhouse gas emissions by 2050 (in relation to 1990 levels). By scenarios we mean combinations of technologies and changes in public behaviour all the way to 2050 that might result in less emissions. The model is quite sophisticated in that it looks at all parts of the British economy and is rooted in scientific and engineering realities – or what is thought to be physically and technically possible. It's also idealistic in the sense that its ultimate aim is to reach the 80% reduction target but not necessarily in the cheapest possible way. Amongst the assumptions of the model is that the level of GDP growth is 2.5% per year.

You can try out this complex model on the web, in the form of the 'pathways calculator'. It allows you to build your own scenarios and see the effects (http://2050-calculator-tool.decc.gov.uk/). Let's look at a few of the scenarios I've built.

Figure 6.4 shows the output of the pathways calculator for a 'business as usual' scenario, in other words a combination of technology and human behaviour that shows little regard for CO_2 emissions. The things that the user can vary (variables) are shown in the list below the three charts. The three charts are the output – or what you get when you press the button to run the model. The variables are quite complex. In the pathways calculator you're allowed four settings, for example in the upper variable on the left – 'domestic transport behaviour' – you're allowed to choose a setting of 1 (in 2050 people travel on average 9% further than they do today and don't much change to public transport) all the way to 4 (people travel the same distances on average and there is a significant shift to public transport). So setting 1 is 'business as usual' and setting 4 is a much greener alternative. In the variable 'CCS power stations' in the second column, setting 1 indicates no increase in CCS up to 2050, while setting 4 envisages 50–90 power stations with CCS.

The result that we get from this business as usual scenario is, as you might imagine, an increase in CO_2 emissions so that Britain would get nowhere near its target of 80% reduction by 2050 (the target is shown as a white dotted line). There's an increase, rather than a steady state, because of the increase in demand as shown by the chart on the top left. Remember that a GDP growth rate of 2.5% per year is fixed in the model.

The next scenario (Figure 6.5) is very green in the sense that the settings for the variables envisage lots of public transport, lots of home

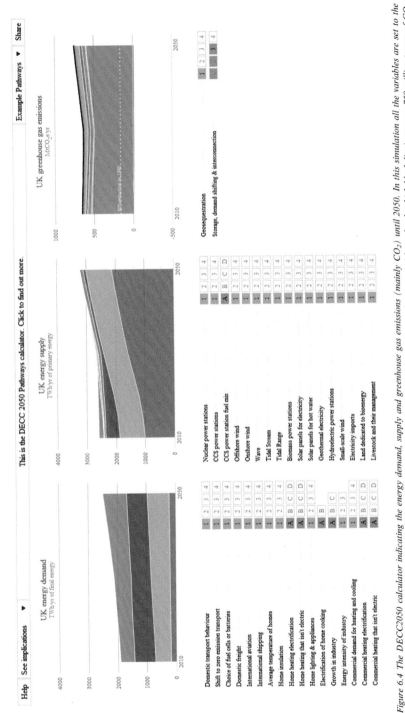

Figure 6.4 The DECC2050 calculator indicating the energy demand, supply and greenhouse gas emissions (mainly CO₂) until 2050. In this simulation all the variables are set to the lowest values – '1' or 'A', so the least effort possible is being spent on reducing CO₂ emissions. The total emissions (shown by the thick black line) are over 750 million tonnes of CO₂ equivalent per year, well above the 2050 target of 160 million tonnes of CO₂ equivalent per year (the white dotted line). The coloured bands in the graphs represent different aspects of demand, supply and emissions. For example in the 'demand graph' the green colour represents energy demand for heating and cooling, while red represents energy demand for transport. Most of the emissions are from fuel burning (transport and electricity from fossil fuels for example). Screenshot from http://2050-calculator-tool.decc.gov.uk/. Note that not all the variables are shown.

Figure 6.5 In this simulation the variables are mostly set to high so this is a very 'green' scenario. For example there are a lot of people using public transport, there is a lot of house insulation — and there is some CCS (from 25 to 40 CCS power stations in Britain). In this scenario, the 2050 target of 160 million tonnes of CO_2 equivalent per year is reached long before 2050. Screenshot from http://2050-calculator-tool.decc.gov.uk/. Note that not all the variables are shown.

insulation, lots of wave and tidal power – and CCS on 25–40 power stations. As you might expect in this scenario, the 2050 target of 160 million tonnes of CO_2 equivalent per year is reached long before 2050. Also it's interesting that the emissions go below the zero line. This is mainly because of CCS. In this scenario by 2030 there are 45 million tonnes of CO_2 being disposed of per year in 2030, 110 million tonnes by 2040 and 177 million tonnes by 2050. Also in the scenario energy demand goes down – because of less need for heating and transport fuel – and supply goes down too. It's obviously important that supply is higher than demand!

The last scenario is 'high CCS' (Figure 6.6) with more power stations than the last scenario. Again the target is reached before the 2050 deadline.

HOW DO YOU MAKE CCS PAY FOR ITSELF?

In the model above, and in Pacala and Socolow's work, the assumption is made that CCS is economically feasible. This is the elephant in the room and probably something that you, the reader, have been asking all the way through this book. CCS is obviously expensive so who pays for it? How does an operating company make a profit? Or are we imagining that governments and treasuries will pay?

You could regard CCS as the reverse of the oil and gas business in that you want to put a gas in the ground forever rather than take it out. You'll probably use a lot of the same equipment that you use in oil and gas, like pipelines and offshore platforms. You'll probably have to do similar geological studies to make sure you're doing it properly. But the fact remains that you're not likely to be extracting CO_2 from the flue gas of a power station to sell it, at least in Europe. In fact you're going to put it into the ground forever. So you can't make any money.

But let's recall the Sleipner example. There the company avoided paying a tax on CO_2 venting to the atmosphere by disposing of the CO_2 permanently underground. The Norwegian government levies a tax on venting CO_2 purely for environmental reasons – to stop a greenhouse gas from doing its worst. The money that would have been spent on tax was diverted to developing the technology and paying for the injection of CO_2 into the Utsira sandstone 1 km below the sea bed.

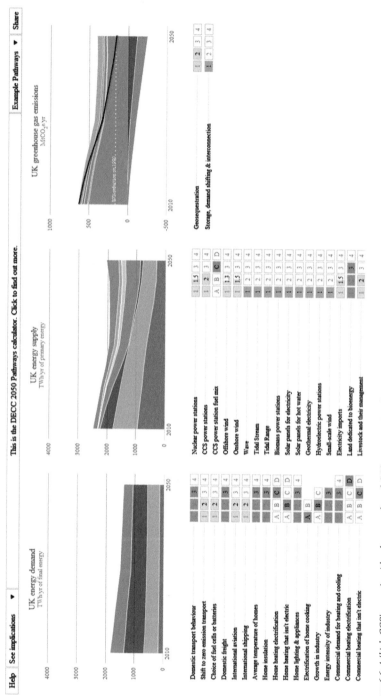

Figure 6.6 A 'high-CCS' scenario with a large number of CCS power stations but assuming less uptake of public transport and less house insulation. Again the 2050 target is reached before 2050. Note that in both the CCS scenarios, CCS causes overall emissions to go below the zero line because CO_2 is actually being permanently removed from the atmosphere, so this counts as 'negative emissions'. Screenshot from http://2050-calculator-tool.decc.gov.uk/. Note that not all the variables are shown.

The European Union Emissions Trading Scheme (ETS) is based on a similar principle, known as 'cap and trade'. The 'cap', set by a government, is a limit to the amount of CO_2 a factory or a power station can emit. This would be set at a figure that would encourage the power station to reduce its emissions rather than put it out of business, but over time it would be expected that the amount of tolerated emissions would be reduced by the government. The companies are given 'permits to emit CO_2' up to the limit, but beyond that they have to buy a further permit (or permits) to emit more. The cost of buying these permits is an incentive to reduce emissions, if the price is high enough. If a power station is able to reduce its emissions below its limit then it can also sell its unused 'permits' to other emitters; this is known as 'carbon trading'. The system is considered a cost-effective way of reducing emissions without the government having to be involved too much. It's a self-sustaining way to reduce emissions because it places a value on a low-CO_2 atmosphere and a penalty on emissions. As permits become scarcer the cost of emission beyond the limit becomes higher because under the law of supply and demand the permits will cost more ... or so the theory goes.

Presently the ETS enforces emissions reductions in 11,000 factories and power stations in 31 countries. These 11,000 point sources are responsible for almost half of the EU's emissions of CO_2. The ETS was set up in 2005 and will run until at least 2020. The caps or limits on emissions for 2020 represent a 21% reduction of greenhouse gases (including CO_2) over the period 2005–2020.

But the price of EU ETS permits has been lower than expected mainly because there is a large surplus. The economic recession of the last few years has resulted in less energy demand so that power stations voluntarily reduced their production under the cap. So the permits are not scarce enough to attract high prices. In fact the cost of emitting carbon at the time of writing is less than €5 per tonne, down from previous levels of over €30 per tonne.

This is what I meant earlier when I said that a cap and trade system can work if the price for emitting carbon is high enough. At the moment it isn't.

What is the price we might aim for? Ideally an operator that wants to do CCS needs to know how much it will cost to do the whole

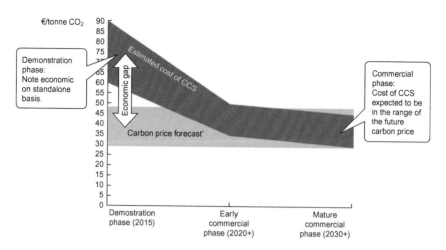

Figure 6.7 *Range of the cost for carbon emission (carbon 'price forecast') and the decline in costs of CCS.* From the Bellona website: http://bellona.org/ccs/introduction-to-ccs/.

process from capture to storage or disposal underground. This figure worked out per tonne of CO_2 captured and stored has to be comparable to the penalty cost of emission. In fact to make money, the penalty cost has to be more than the cost of CCS. When the figure is more, then CCS will be a viable business.

Of course this isn't as easy as it sounds. The penalty cost for carbon emission is hard to predict. When the ETS was introduced no one would have predicted that it would fall to less than €5 per tonne. Analysts think that it might eventually stabilize at between €30 and €50 per tonne. Recent studies suggest that the first CCS projects in the power sector will cost between €60 and €90 per tonne of CO_2 captured and stored. But these costs are likely to come down as the industry gets better at the technology. The costs are expected to go down to between €35 and €50 per tonne in the early 2020s.

We can illustrate this in a simple graph (Figure 6.7) which shows the range of the cost for carbon emission (carbon 'price forecast') and the decline in cost of CCS. As the graph shows, we're in a period where CCS is 'not economic on a standalone basis', in other words it isn't a viable business. There may later be a 'commercial phase' where the cost of CCS is similar or less than the 'carbon price'. Then the business will be viable and profitable.

So while CCS isn't economic, how do we stimulate it? How do we make sure that the technology works together in a chain at the big scale we need it? Why would a company, even a very profitable company, spend a lot of money doing something it won't make money out of for several years?

Well it won't, but it might try if some of its risks can be reduced, mainly the financial risk of trying something very new. Because some governments are convinced that CCS has to be part of an overall CO_2 reduction strategy (one of Pacala and Socolow's wedges), they are prepared to support companies financially to help them develop the technology. An example is the UK CCS Commercialisation Competition which has made up to £1 billion of government money available to encourage practical experience in the design, construction and operation of commercial-scale CCS. The competition was open to any company that wished to develop a commercial-CCS project in Britain. The current competition opened in April 2012, and closed in July 2012 and four full chain projects were shortlisted. In March 2013 this was narrowed down to two projects, both of which might receive money.

Engineers consider that this period of public-funded demonstration is vital to CCS because the industry has to learn from the mistakes that it will inevitably make. The early demonstration projects ('demo' projects) are the test beds out of which the full chain technology will emerge. They will also start the process of cost reduction that is needed to make CCS viable as a business. Engineers say that the projects might overlap so that learning from one project feeds into the next so that the overall progression to a viable commercial CCS business ('global rollout') is accelerated (Figure 6.8).

Predicting when global rollout will happen is obviously very difficult but one suggested programme is shown in Figure 6.9. The article in which this prediction was presented was published in 2008 and will no doubt be considered by many to be too optimistic, but it suggests that demonstration projects will be operating by 2015 and that leading countries will have commercial CCS by 2020. According to this forecast CCS will be a global business by 2025 but whether a 'wedge' will be achieved will depend on how many power stations are operating with CCS.

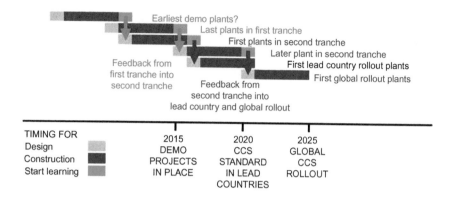

Figure 6.8 CCS projects might overlap in time so that learning from one project feeds into the next so that the overall process arriving at a viable commercial CCS business (global rollout) is accelerated. From Gibbins and Chalmers (2008).

ISN'T CCS ARTIFICIAL?

At a recent conference I heard an environmentalist who was not convinced by its feasibility say that CCS is like 'liposuction on an obese person'. This is a colourful and interesting comparison because it encapsulates the ambiguity that some environmentalists feel about CCS. They see it as an unnecessarily complex solution to a simple problem – and possibly a solution that doesn't work. This sounds like a paradox, but what many environmentalists feel is that CCS allows the fossil fuel business to continue. It lets the big fossil fuel companies – the oil companies, the coal companies – off the hook. It doesn't force them to change the fundamental way that they operate. Put in another way CCS is a technological solution to a societal problem. Human societies have bad habits: they use too much energy; they waste too much energy; they don't value the environment. CCS allows us to continue as before. The obese person continues to eat too much and doesn't address the real problem.

Let's look at this point by point. How artificial is the industrial process of burying a waste material underground? In fact underground disposal is a big part of many industries and it is going on all around us quite routinely. In the United States, for example, there are over 144,000 disposal wells which inject billions of litres of water every day into deep rock layers. Known as Class II wells according to the US Environmental Protection Agency (EPA), they inject 'fluids associated with oil and natural gas production'. This fluid is mostly saline water from deep reservoirs

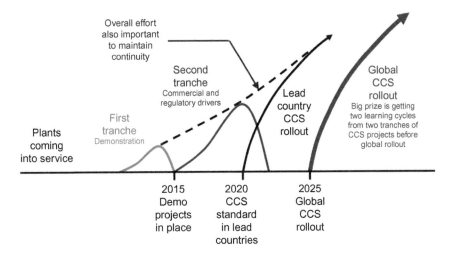

Figure 6.9 The global rollout of CCS. From Gibbins and Chalmers (2008).

that comes up along with oil or gas, when it's produced. This amount of very salty water can't be put into rivers or lakes and so deep disposal is necessary. Class I wells, of which there are over 500 in the United States, are used for more hazardous liquid waste from petroleum refining, pharmaceuticals, food production and other sources.

With the increase of extraction of shale gas where very large amounts of water are used for hydraulic fracturing, a lot of water has to be disposed off also. At present with a boom in shale gas extraction in Texas, 290 million barrels of wastewater – equivalent to about 18,500 Olympic-size swimming pools – are being disposed of underground each month. Of course we use the subsurface for storing water and gas as well. There are over 120 underground storage facilities in Europe for natural gas and these can be reused indefinitely, coping with very large volumes of gas.

It's true that CCS doesn't address the problem of energy waste, nor encourage people to think about how they consume. It doesn't encourage the development of renewable energy either. But a glance at the forecasts for coal use in the next 50 years in India, China and South Africa (see Chapter 1) shows how committed developing countries are to using their coal. It's clear also that the coal will be used for electricity generation in ever larger power stations. At the moment there's only one technology that can deal with the emissions of CO_2 that would almost certainly result from this increased use of coal and that is CCS.

More recently energy experts have been talking about a 'dash for gas'. More gas is being found worldwide particularly in shale. In the United States almost a third of domestic gas comes from shale, up from almost nothing a decade ago. Could the world be powered by gas in the next few decades? There is a climate advantage to gas: it burns cleaner than coal because it produces less CO_2 and other pollutants. To countries like Poland — which depends almost entirely on coal for its electricity — switching from coal to gas to generate electricity would be convenient if it could produce its own gas from shale, and it would reduce Poland's CO_2 emissions. Home-grown shale gas would also help Poland to be independent of gas suppliers on which it currently depends.

If the world moves more to gas than the IEA and other institutions predict, then CCS could be made to work with gas power stations in the same way as it does with coal. This book has been mainly about CCS and coal, but it could just as easily be written for a gas future.

What about the charge that CCS lets fossil fuel companies off the hook? Does CCS mean that we commit too much to coal and gas, and continue our fossil fuel addiction? Interestingly this concern turns up amongst CCS experts: the idea of 'lock-in' to 'unabated' coal. What CCS experts worry about is that the promise or the prospect of CCS will encourage companies to build coal and gas power stations while assuming that these stations can easily be fitted with CCS after they've been built ('retrofitted'). As we've seen earlier this isn't simple. Power stations can't be built just anywhere for CCS. A major consideration is whether there's a CO_2 pipeline nearby; even more important, a storage site needs to be accessible. Many advocates of CCS mention the need for 'capture-ready' power stations to be built — in other words power stations that can be easily be adapted for CCS. If this isn't done then power stations of the old kind will continue to pollute. And since the average life of power stations is many decades this is a serious problem — the problem of 'lock-in'.

Most CCS scientists see the technology as a 'bridge' to a time when truly renewable sources of energy and electricity can take up the strain of global power demand. So they don't want 'lock-in' to coal or CCS either, only to run the technology until such time as it isn't needed anymore. Renewables can't at the moment satisfy our demand for energy. As I write I'm checking an app on my iPad called 'UK Energy' which monitors the sources of electricity generated in Britain minute by minute. At this moment 42.7% of Britain's electricity is coming from coal, and only 7.5% from wind and Britain is a country that shows exceptional commitment to the development of renewables. It's also a windy place!

I've shown that our forests and our seas capture CO_2 as part of natural processes; otherwise we wouldn't have the enormous deposits of coal and oil of planet Earth. The formation of limestone on the seabed — which is happening now in our tropical seas and which happened in huge amounts over geological time — is not unlike the process that might happen in advanced CCS where CO_2 reacts with pore water and rock particles in the reservoir. The carbonate minerals formed in both processes are similar. But natural processes like coal and limestone formation don't capture CO_2 at the rate that we need to capture it. CCS can capture at the right rate to make a difference and processes of the carbon cycle are operating in CCS, so it's not so artificial. It's just a bit faster.

BACK TO THE CARBON CYCLE

Let's look at a nightmare scenario now. I don't mean to frighten you at the end of this book — after all it's much better to end on a high note — but I want to show you what can happen when CO_2 in the atmosphere increases very fast. We have to return to the carbon cycle which I touched on earlier in the book to understand what actually happened 55 million years ago. This event, known as the Palaeocene–Eocene Thermal Maximum (PETM), shows how earth can change fast, how life in the oceans and on land can die back but also how it can recover. Studying geological history — whether it be glaciation in the Carboniferous period or tropical water temperatures at the pole in the PETM — shows us that life on Earth is resilient over long time scales. Earth recovers from extinctions and mass die-offs but it isn't much fun if you're the species that's dying off.

So what happened at the PETM? To understand this it's best to return to the rocks. Most people think rocks are pretty boring. CCS or oil geologists and engineers like them better because they can act as containers or caprocks. But other kinds of geologists — palaeontologists and stratigraphers — look on rocks quite differently. Sedimentary rocks like shale — which you might remember I mentioned as being a good caprock — are also very useful in that they record how the environment changes over time. The thin layers of shale — and they can be very thin, a tenth of a millimetre isn't unusual — *record* aspects of the environment in which they're formed. The most obvious thing they tell us is the kind of sediment that was deposited, but they also contain fossils of once-living organisms and chemical traces of the ancient environment. Mostly we think of fossils as faint impressions of shells or ammonites on the surfaces of rocks, but fossils can also be very small — too small to see — for example fossil spores or pollen, or fossil plankton. The shale can sometimes contain thousands of fossils of this type in a single thin fragment. The organic matter that's preserved amongst the particles of shale also has a value in that it might record isotopic ratios of the ancient environment, for example $\delta^{13}C$.

Each thin layer represents a single episode of sedimentation on the sea bed or a lake bed, and if you break open the shale, splitting it along its layers, the surfaces you expose were once a little part of an ancient sea bed. In fact the layers of shale seen side-on sometimes look like the pages of a book and those pages can be 'read' or interpreted like a book if you collect the fossils and measure the $\delta^{13}C$. The 'pages' tell you how the environment changed through time. Of course the lower you go in the shale layers, the further you're going back in time.

An example of an interpreted set of shale layers is shown in Figure 6.10. The shale was drilled through while looking for oil in the central part of the North Sea and a cylindrical core of the rock was brought up from below for company geologists and scientists to study. Although this core of rock is from the North Sea, similar patterns are found around the world at this time. Originally the core of rock would have been used to understand the best place to look for oil or gas, but it was passed on to palaeontologists and stratigraphers to study the PETM.

The diagram shows the depth in the borehole on the left, increasing downward. You can think of this as being related to time. The lower you go in the borehole the older the rock is. If you study the fossils and chemistry of the rock in the core going *upward* then you're following the progress of time, seeing the changes as they occurred. The part to concentrate on is on the left. This shows the $\delta^{13}C$ of the total organic carbon (TOC) in the shale. The PETM is marked by the sudden spiky jump to the left that begins at about 2614 m in the borehole (called 'CIE onset' in the diagram). This is an extraordinarily large

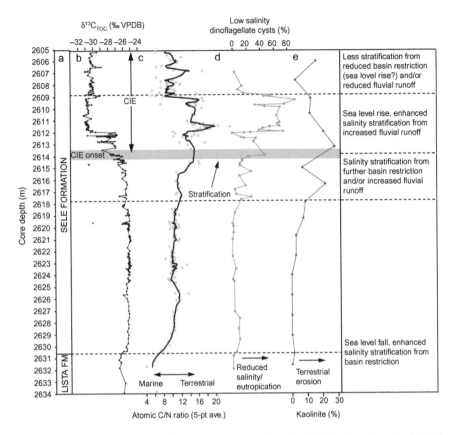

Figure 6.10 The pages of the 'shale book': an interpreted record from the PETM in the rocks under the North Sea. The PETM starts at a depth of about 2614 m in the borehole. In the diagram it's marked as 'CIE' which stands for carbon isotope excursion. From Kender et al. (2012).

jump which means that the atmosphere and the carbon cycle must have been changing a lot and very rapidly. More about this later.

Elsewhere rock cores of the PETM have revealed fossil shells that allow us to see other changes in isotopic ratio. For example the progression of $\delta^{18}O$ through the time of the PETM tells us something about the temperature of seawater. The startling results are that for a very short period of time in Earth's history (less than 200,000 years), sea surface temperatures rose by between 5°C and 8°C but then decreased again. This spike in temperature is shown in Figure 6.11 (top left of the graph) and the 200,000 year period is so thin that it looks like a single line on the scale of million-year time intervals. The temperature graph suggests that the Earth's oceans reached their

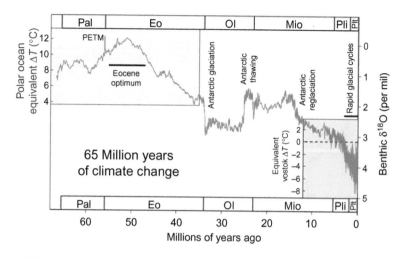

Figure 6.11 The temperature spike (in the green line) of the PETM from $\delta^{18}O$ of fossil shells. From Wikipedia
https://en.wikipedia.org/wiki/Paleocene%E2%80%93Eocene_Thermal_Maximum.

warmest for 65 million years during this short time period. It's perhaps the most extreme period of global warming in Earth history. So far....

It's known that the sea became more acidic during the PETM, and that there were lots of extinctions in the Earth's seas, either because of the decrease in pH or the warmer temperatures. Many ocean plankton for example simply couldn't cope with the changes. The sea level rose as well because of ice melting but also perhaps because of thermal expansion of seawater. We're not sure what the effects were on land because the PETM is best represented in marine sediments but some areas give us a feel for what might have happened.

The North Sea for example looked rather different during the PETM, 55 million years ago (Figure 6.12). It was more enclosed by land, except to the north. So there was probably no English Channel at the time. The Atlantic Ocean was much narrower so that Greenland and Scandinavia were almost touching. Iceland didn't even exist yet.

Because the North Sea was more enclosed, the sediments deposited during the PETM have more information about the changes on the land. If you go back to Figure 6.10 you'll see some other vertical graph lines, for example the green line showing the percentage levels of 'low-salinity dinoflagellate cysts'. Dinoflagellate cysts are the fossils of a particular kind of plankton and the ones that were measured for the graph

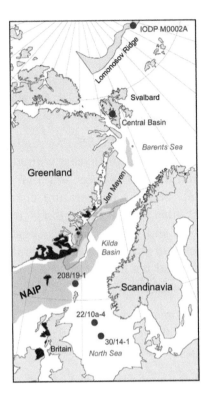

Figure 6.12 The North Sea was more enclosed by land 55 million years ago. Some of the effects of the PETM on land can be seen in the sea bottom sediments deposited at the time. From Kender et al. (2012).

were ones that we know preferred low-salinity seawater. What their increase seems to suggest is that the surface waters of the North Sea became less salty during the PETM. This might have been because there was suddenly a lot more rainfall on the land around and this was brought by rivers making the sea less salty. But to cause such a huge salinity change in a large sea implies an enormous input of freshwater from the land. The spores and pollen from land plants around the North Sea which we find in the sediment layers at this point also suggest turmoil with whole ecologies substituting one for another as the sea level and temperature rose. These are just a few of the environmental changes that have been interpreted from the data, and again we're not sure of the effect this would have had on wider ecology or on a society like our own, if it had existed at the time. But it's likely that the changes were all caused directly or indirectly by global warming.

This is where it gets scary though. How does the rate of change through the PETM compare with forecasts for model global warming?

You might remember from earlier in the book that the most ambitious of the plans for emissions reduction want to ensure the limit of temperature increase to 2°C in the next century. You might also recall modern $\delta^{13}C$ records of the carbon in atmospheric CO_2 from Mauna Loa Observatory (Figure 1.2) which show a drop of about 1‰ in the last 20 years. This drop is related to the addition of CO_2 from fossil fuels to the atmosphere. ...

Well the most recent work on the PETM suggests two steps in the drop of $\delta^{13}C$, each lasting about 1000 years with about 20,000 years in between, but with a full drop of between 2‰ and 7‰ over that 22,000 year period. The total seawater temperature change suggested from oxygen isotopes and other chemical methods over that very long period was between 5°C and 8°C. What this suggests then is that the *present* rate of change is *already* much faster than seems to have happened over the PETM, and it's likely to speed up. The PETM caused rapid environmental change: what can we expect?

Temperatures did become more normal after the PETM but this took between 30,000 and 150,000 years. Big global carbon storage processes probably rectified things but not fast enough to save a quite a few species.

For those of you that are curious about why the PETM happened, it appears that greenhouses gases (either CO_2 or methane) were being added to the atmosphere at a very rapid rate, rather like they are now. No one's really sure where these gases came from. Some believe that methane from sediments in the sea bed was released in a great oceanic burp; others think that big volcanoes suddenly released a huge amount of CO_2.

Someone once said that *history is bunk*, implying that there's no point in looking back. Probably whoever said it wasn't thinking of geological history, but the principle still applies. I prefer Edmund Burke's 'Those who don't know history are destined to repeat it'. The value of geology – of palaeontology and stratigraphy – is that it allows us to see what's happened to the planet in the deep and distant past. The view we get is hazy like a faded photograph – and it's hard to interpret – but we shouldn't ignore the warnings.

FULL CIRCLE

In this book I've tried to show how the carbon cycle – the never ending exchange of carbon between the earth, the biosphere, and the atmosphere — results in fossil fuels like coal. You might look upon coal as a solid reminder of the CO_2 of an ancient atmosphere. Coal and other fossil fuels have been used by mankind for thousands of years but when they're burnt they return the carbon that was naturally captured back into the atmosphere so that CO_2 concentrations increase. So you get global warming with unpredictable results. Of course the very creation of these fossil fuels can cause climate change because of a 'reverse greenhouse effect' — so the whole thing works as a huge cycle.

On planet Earth we want to preserve our lifestyle but also live within our limits. We all know that we can't spend more than we earn. The environmental movement has shown us that we ought to value the environment and that we shouldn't 'spend' our ecological heritage unwisely. You could argue that that's exactly what we've done since the industrial revolution. We should develop renewable sources of energy while weaning ourselves off fossil fuels but we need to do this at a pace that allows us to keep the lights on.

CCS might help us to do that.

BIBLIOGRAPHY

Bellona website. <http://bellona.org/ccs/>.

CCSA website. <http://www.ccsassociation.org/>.

DECC pathways calculator. <https://www.gov.uk/2050-pathways-analysis>.

DECC website. <https://www.gov.uk/uk-carbon-capture-and-storage-government-funding-and-support>.

Gibbins, J., Chalmers, H., 2008. Preparing for global rollout: a 'developed country first' demonstration programme for rapid CCS deployment. Energy Policy 36, 501–507.

Kender, S., Stephenson, M.H., Riding, J.B., Leng, M.J., Knox, R.W.O'B., Vane, C.H., et al., 2012. Enhanced precipitation and vegetation changes in the North-East Atlantic at the Palaeocene–Eocene boundary. Earth Planet. Sci. Lett. 353–354, 108–120.

Pacala, S., Socolow, R., 2004. Stabilization wedges: solving the climate problem for the next 50 years with current technologies. Science 305, 968–972.

Sluijs, A., Bowen, G.J., Brinkhuis, H., Lourens, L.J., Thomas, E., 2007. The Palaeocene-Eocene thermal maximum super greenhouse: biotic and geochemical signatures, age models and mechanisms of global change. In: Williams, M., Haywood, A.M., Gregory, F.J., Schmidt, D.N. (Eds.), Deep Time Perspectives on Climate Change: Marrying the Signal from Computer Models and Biological Proxies. The Geological Society, London.

US EPA website. <http://water.epa.gov/type/groundwater/uic/class2/index.cfm>.